Elementary thermodynamics for geologists

Elementary thermodynamics for geologists

B. J. WOOD

AND

D. G. FRASER

OXFORD UNIVERSITY PRESS

1976

Oxford University Press, Walton Street, Oxford OX2 6DP

OXFORD LONDON GLASGOW NEW YORK
TORONTO MELBOURNE WELLINGTON CAPE TOWN
IBADAN NAIROBI DAR ES SALAAM LUSAKA ADDIS ABABA
KUALA LUMPUR SINGAPORE JAKARTA HONG KONG TOKYO
DELHI BOMBAY CALCUTTA MADRAS KARACHI

Casebound ISBN 0 19 859926 9
Paperback ISBN 0 19 859927 7

© Oxford University Press 1976

Printed in Great Britain
by Thomson Litho Ltd., East Kilbride

Preface

This is not intended to be a conventional thermodynamics
text book. Its aim is to introduce the uninitiated
reader to a rapidly growing branch of geology without a
lengthy initial treatment of classical thermodynamics.
To do this we have tended to employ operational
definitions where possible rather than to develop the
logical coherence of the subject. This treatment clearly
tends to underplay the intrinsic beauty of thermodynamics
and its ability to generate new relationships by
manipulating the equations of its structure. While this
may offend the more knowledgeable reader, we believe
that it is a very effective way of introducing
thermodynamics to geologists. In this respect
thermodynamics is like a natural language. It is a
simple language in that it has a small vocabulary (about
fifty or so words) and a rigorously defined grammar. Thus
only a strictly limited number of valid sentences, or
equations, can be formed. In this book we shall be
concerned not with the grammar of thermodynamics but with
how to read, speak and understand it as a living language.

PREFACE

We hope that the reader will be able to use this book as a "work-shop manual" for the subject and that he will be sufficiently stimulated to try to apply the methods to geological problems of his own interest. If this aim is successful, he should discover some of the inherent power of the subject and will rapidly uncover many of its pitfalls and limitations.

D.G. Fraser, Oxford
B.J. Wood, Manchester

August 1976

Acknowledgements

The information and ideas presented in this book are in
large part the products of discussions with our colleagues.
In this context we wish particularly to acknowledge Drs.
J.S. Anderson, S. Banno, I.S.E. Carmichael, W.S. MacKenzie,
R.K. O'Nions, R. Powell and E.J.W. Whittaker, from whom we
have learnt much. We are particularly grateful to Dr.
Whittaker for his care in reading the manuscript. The
substance of the book was tested on several groups of
students and in particular during graduate courses in
Theoretical Petrology held in Manchester. We thank the
students for their discussion and criticism of the course
material and for their forbearance. We also acknowledge
the financial support of the Natural Environment Research
Council which enabled us to stage the graduate courses.

We thank Mike Brayley and Peter Deussen for their help
with diagrams, Sue Maher for preparing photographs, and
Pat Jackson for assistance with typing.

Finally, anyone who has experience of trying to produce
immaculate typescript containing formulae with multiple
subscripts and superscripts in two alphabets will appreciate
how much we owe thanks to our typist, Patricia Crook.

List of Contents

LIST OF CONTENTS

x

LIST OF CONTENTS

LIST OF CONTENTS

LIST OF CONTENTS

Units

A considerable amount of thought and discussion went into
the seemingly trivial matter of the choice of units to be
used in this book. Our problem arose from the fact that
most chemists, physicists, and engineers now use the SI
system of units. Most undergraduate students have also
been taught SI units at school. SI units are, however,
little used in the geological literature and many papers
containing c.g.s. and other non-SI units continue to be
published. In particular, the tables of thermodynamic
data, on which many of the calculations described here
depend, use calories (cal) instead of Joules (J) as units
of energy. Pressures are rarely given in Newton m^{-2} (Pa)
but almost always in bars or kilobars (kbar); 1 bar =
10^5 N m^{-2}.

In order to remain consistent with the literature we
have, with some reluctance, opted to retain calories and bars
as our main units of energy and pressure. Wherever
convenient we also give the SI equivalents in the text. As
an example of the use of the different systems of units,
we may consider the following.

Example. The free energy of a solid phase $b(G_b)$ at some
pressure P and temperature T is given approximately by
(from eqns (1.42 - 1.44) :

where G_b $\quad=\quad$ $H_{1,T} - TS_T + (P-1)$ V

$H_{1,T}$ $\quad=\quad$ Enthalpy of b at 1 bar and T,

S_T $\quad=\quad$ Entropy of b at T,

V $\quad=\quad$ Volume of b.

Let us calculate G_b per mole of b at 800 K and 10 kbar given :

$H_{1,800}$ = 15 000 cal mol^{-1} = 15.0 kcal mol^{-1}

$\qquad$ (cal mol^{-1} = calories per mole)

S_{800} = 10 cal mol^{-1} deg^{-1} = 10 entropy units (e.u.)

V $\quad=$ 5 cm^3 $=$ $\frac{5}{41.84}$ = 0.1195 cal bar^{-1} (calories
$\qquad\qquad\qquad\qquad\qquad\qquad\qquad\qquad\qquad$ per bar)

Note that to be consistent, volumes must be expressed
in units of calories per bar.

$\qquad$ At 10 kbar = 10 000 bar and 800 K,

G_b $\quad=$ 15 000 - 800 (10) + 9999 (0.1195)

$\quad=$ <u>8195 cal mol^{-1}.</u>

In SI units $H_{1,800}$ is in J and S in JK^{-1} mol^{-1}
(Joules per degree K per mole). V is expressed in cubic
metres (m^3). Using the conversion factors :

$$1 \text{ cal} = 4.184 \text{ J}$$

$$1 \text{ cm}^3 = 10^{-6} \text{ m}^3$$

$$1 \text{ bar} = 10^5 \text{ N m}^{-2} \text{ (Newtons per square metre)}$$

the corresponding values are :

$$H_{1,800} = 62760 \text{ J mol}^{-1}$$

$$S_{800} = 41.84 \text{ JK}^{-1} \text{ mol}^{-1}$$

$$V = 5 \times 10^{-6} \text{ m}^3$$

At 10 000 bar (10^9 N m^{-2}) and 800 K,

$$G_b = 62760 - 800 (41.84) + (10^9 - 10^5) 5 \times 10^{-6}$$

$$= \underline{34288 \text{ J mol}^{-1}}$$

Using kiloJoules (kJ) or kilocalories (kcal), we have

$$G_b = \underline{34.288 \text{ kJ mol}^{-1} = 8.195 \text{ kcal mol}^{-1}.}$$

I. Introduction and definitions

The mineralogies of rocks sampled at the earth's surface reflect the pressures and temperatures at which the rocks were formed and their chemical compositions. The methods of chemical thermodynamics can be used to predict the most stable, or equilibrium, mineral assemblage of a rock under any given pressure-temperature conditions.

It has long been apparent that chemical equilibrium is approached at some time during the crystallization history of many metamorphic (e.g. Eskola 1920) and slowly cooled igneous rocks and even during rapid cooling of volcanic rocks (e.g. Carmichael 1967). The application of thermodynamic principles to these rocks can therefore provide a valuable insight into their histories.

In such cases it may be valid to assume equilibrium and to apply thermodynamic considerations in order to determine physical conditions (pressure, temperature, etc.) of crystallization. It is apparent that the volume of rock which is in equilibrium may be small because of the preservation of original compositional differences on the cm scale even during high-grade metamorphism.

Given the fundamental assumption of chemical equilibrium we shall begin by defining the important thermodynamic terms and variables.

1.1 SYSTEM

A system is simply a group of atoms, minerals, or rocks
which are under consideration. The exact position of the
boundary of a system may be fixed at will. For example, it
may be convenient to choose as the system a whole outcrop;
or a hand specimen; or just one mineral in a thin section.
The boundary of a system is generally fixed in such a way
that the minerals, fluids and melts within it may be regarded
as being in equilibrium.

Changes that take place within a system may or may not
involve interaction with surrounding material. An isolated
system is one which cannot exchange energy or matter with its
surroundings. A closed system can exchange energy but not
matter with its surroundings, e.g. isochemical metamorphism.
An open system may exchange both energy and matter with its
surroundings, e.g. metamorphism with accompanying metasoma-
tism.

A system is made up of one or more phases. A phase is
a restricted part of the system with distinct physical and
chemical properties. When considering rocks one often
identifies several mineral phases, e.g. all the olivine
crystals in the system constitute the olivine phase, the
plagioclase crystals, the plagioclase phase and so on. In
addition there may be a molten phase and a fluid phase
containing, for example, H_2O, CO_2 and CO.

1.2 COMPONENTS

Each phase in a system may be considered to be composed of one or more components. For any particular phase these components may be defined in a number of different ways. For example, in an $(Mg,Fe)_2SiO_4$ olivine solid solution some of the possible sets of components are : (a) Mg_2SiO_4 and Fe_2SiO_4, (b) MgO, FeO, and SiO_2, (c) Mg^{2+}, Fe^{2+}, Si^{4+}, and O^{2-}. The choice of components is arbitrary and depends on the type of thermodynamic problem which is under considera-tion. Generally, however, it is most convenient to take the end-members of solid solution series as the components of complex mineral phases, e.g.

Mg_2SiO_4 and Fe_2SiO_4 in $(Mg,Fe)_2SiO_4$ olivine,

$KAlSi_3O_8$, $CaAl_2Si_2O_8$ and $NaAlSi_3O_8$ in

$(K,Na,Ca)(Al,Si)_2Si_2O_8$ feldspar

$CaMgSi_2O_6$, $CaFe^{2+}Si_2O_6$ and $NaAlSi_2O_6$ in

$(Ca,Na)(Mg,Fe^{2+},Al^{3+})Si_2O_6$ clinopyroxene

The reason for the choice of these end-member formula units is that thermodynamic data are not available for all possible choices of component. For example, the thermodyna-mic properties of pure Mg_2SiO_4 olivine and $NaAlSi_3O_8$ plagioclase are known, but those of Mg^{2+} in olivine and Na^+ in plagioclase are not. This is because the latter cannot be studied in isolation.

1.3 CHEMICAL POTENTIAL, μ

In order to determine the directions of chemical change in rocks, it is necessary to introduce the concept of the chemical potentials (μs) of the components present in the rocks. A chemical potential is analogous to an electrical or gravitational potential in that by measuring the μs of components in different parts of the system, the tendency (or otherwise) of material to flow or react can be deduced. The direction of flow of components is always from regions of high chemical potential to those of low chemical potential, so that the system as a whole approaches its lowest possible energy state.

Consider a system consisting of two polymorphs $CaCO_3$, pure aragonite ($CaCO_3$), and pure calcite ($CaCO_3$). The chemical potential of the $CaCO_3$ component in the calcite

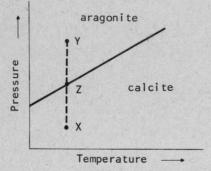

FIG.1.1. General form of the boundary between calcite and aragonite in the $CaCO_3$ system.

phase ($\mu_{CaCO_3}^{calcite}$) relative to that in the aragonite phase ($\mu_{CaCO_3}^{aragonite}$) at different pressures and temperatures may be deduced from the calcite-aragonite phase diagram (Fig. 1.1).

At the point X, which is in the calcite stability field, $\mu_{CaCO_3}^{calcite}$ is less than $\mu_{CaCO_3}^{aragonite}$ and material will flow from the aragonite phase to the calcite phase until all the aragonite disappears. At Y, the reverse will happen, $\mu_{CaCO_3}^{aragonite}$ being lower than $\mu_{CaCO_3}^{calcite}$. At Z the values of μ_{CaCO_3} are <u>equal</u> in the two phases; thus there will be no net flow of material and the two phases will be in equilibrium (see Fig. 1.2).

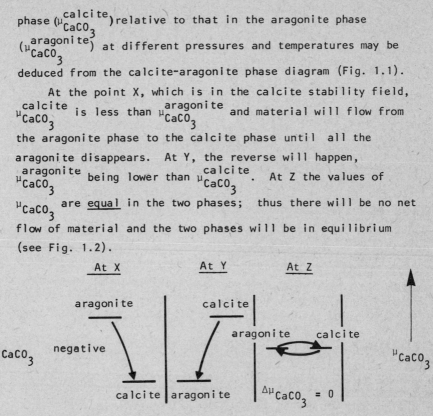

FIG. 1.2. Relative chemical potentials at the points X, Y, and Z in Fig. 1.1.

To express the relationship between calcite and aragonite it is customary to write a balanced chemical reaction in terms of the component (or components) of interest and to consider the relative values of μ in the manner discussed above :

$$CaCO_3 \rightarrow CaCO_3. \qquad (1.1)$$
$$\text{aragonite} \quad \text{calcite}$$

By convention the reaction is considered to proceed from left to right, so that the components on the right-hand side of the reaction are taken to be products and those on the left-hand side the reactants. Using this convention the difference in μ_{CaCO_3} between products and reactants, $\Delta\mu_{CaCO_3}$ is defined as follows :

$$\Delta\mu_{CaCO_3} = \mu_{CaCO_3}^{products} - \mu_{CaCO_3}^{reactants}$$

$$\Delta\mu_{CaCO_3} = \mu_{CaCO_3}^{calcite} - \mu_{CaCO_3}^{aragonite}. \qquad (1.2)$$

Thus, in Fig. 1.1 above :

(a) At X, $\Delta\mu_{CaCO_3} < 0$,

(b) At Y, $\Delta\mu_{CaCO_3} > 0$,   (1.3)

(c) At Z, $\Delta\mu_{CaCO_3} = 0$.

The fundamental condition for chemical equilibrium is expressed by eqn (1.3c).

The chemical potentials of components are expressed in joules/mole (J mol^{-1}) or, more commonly in the geological literature, calories/mole (cal mol^{-1}). Since chemical potential is a molar property it is independent of the actual number of moles present in the system, i.e. $\mu_{CaCO_3}^{aragonite}$ is fixed for any given pressure and temperature regardless of whether there are 1 or 1000 moles of $CaCO_3$ in the system.

(An analogy may be drawn with the gram formula weight of a component. The component $CaAl_2Si_2O_8$ has a formula weight of 278.21 and a gram formula weight of 278.21 grams. The gram formula weight (a molar property) is independent of whether there are 2 moles, 10 moles or 10^6 moles of $CaAl_2Si_2O_8$ in the system.)

Properties of the system like μ which are independent of its mass are known as <u>intensive properties</u>. Some other examples are pressure, temperature, and refractive index.

Chemical potential is a function of state of the system. This means that the chemical potential of any component i, μ_i, has a unique value if pressure, temperature, and bulk composition of the system are fixed. As an example let us go back to Fig. 1.1. If we take a system consisting of pure $CaCO_3$, then at the point X, $\mu_{CaCO_3}^{calcite}$ and $\mu_{CaCO_3}^{aragonite}$ have unique values. These values are completely independent of how the system reached the pressure and temperature corresponding to X. The only way in which $\mu_{CaCO_3}^{calcite}$ or $\mu_{CaCO_3}^{aragonite}$ can be changed is by a change in the <u>state</u> of the system, i.e. by changing pressure or temperature or by adding other components.

1.4 GIBBS ENERGY AND CHEMICAL POTENTIAL

In the general case of phases and systems consisting of more than one component, it often becomes necessary to consider the thermodynamic properties of the phase (system) as a whole rather than the properties of its constituent components. The sum of the chemical potentials of each component times the number of moles of each for all the components in a phase or system is defined as the Gibbs energy (or free energy) :

$$G_{total} = \sum_i \mu_i n_i. \tag{1.4}$$

where μ_i is the chemical potential of component i, n_i is the number of moles of component i in the phase of interest, and the summation extends over all components. For an olivine solid solution consisting of n_1 moles of Mg_2SiO_4 and n_2 moles of Fe_2SiO_4, the free energy of the olivine phase is given by

$$G_{total} = n_1\, \mu^{ol}_{Mg_2SiO_4} + n_2\, \mu^{ol}_{Fe_2SiO_4}. \tag{1.5}$$

It is apparent from eqns (1.4) and (1.5) that G is a function of the number of moles of components present in the phase or system of interest. Properties of the system which depend on the amount of material present are called <u>extensive properties</u> (e.g. mass, volume, heat content).

Since the free energy of a multicomponent phase is a function of its composition, it is apparent from eqn (1.4) that the chemical potential of any component i in the phase may be obtained by the partial differentiation of G with respect to n_i for constant amounts of all other components (n_j) :

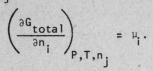

$$\left(\frac{\partial G_{total}}{\partial n_i}\right)_{P,T,n_j} = \mu_i. \tag{1.6}$$

The subscript variables P (pressure), T (temperature), and

numbers of moles of components other than i (n_j) are all
held constant during the differentiation.

For a pure one-component phase the following relationship is apparent from eqn (1.5) :

$$G = \mu n$$
$$\mu = G/n. \tag{1.7}$$

So for a pure phase the chemical potential of its constituent
component is equal to the molar free energy of the phase.
From eqn (1.7) it can be seen that, since the units of μ are
cal mol^{-1} (J mol^{-1}), the units of G are simply calories
(Joules).

Throughout the rest of this book it will be convenient
to deal almost exclusively with <u>molar</u> properties of phases
or of components of phases. For this reason we will use G_a,
unless otherwise stated, to refer to the free energy per
mole of phase a and μ_i^a to refer to the chemical potential of
component i in phase a. If a is pure component i we
therefore have

$$G_a = \mu_i^a. \tag{1.8}$$

Absolute values of free energies of phases, or of
chemical potentials of components, cannot and need not be
determined. The free energies of phases can, however, be
measured relative to an arbitrary set of standard substances,
usually either elements or oxides, so as to construct a scale
of relative free energies or chemical potentials. Since the
equilibrium condition for mass-balanced reactions (equal

numbers of atoms on both sides) is expressed as a zero difference
in chemical potentials, one needs only to know the difference
between μs of reactant and product components in order to
determine whether the reaction should proceed from right to left
or vice versa.

1.5 EQUILIBRIA INVOLVING MORE THAN ONE COMPONENT

Equilibria in systems containing more than one component
can be considered in a similar manner to that already
discussed for the aragonite-calcite reaction. At high
pressures, anorthite decomposes according to the reaction

$$3CaAl_2Si_2O_8 \rightleftharpoons Ca_3Al_2Si_3O_{12} + 2Al_2SiO_5 + SiO_2.$$
$$\text{plagioclase} \qquad \text{garnet} \qquad \text{kyanite} \qquad \text{quartz} \qquad (1.9)$$

By analogy with eqns (1.2) and (1.3), ΔG (more than one
component involved) may be defined as follows :

$$\Delta G = \mu^{garnet}_{Ca_3Al_2Si_3O_{12}} + 2\mu^{kyanite}_{Al_2SiO_5} + \mu^{quartz}_{SiO_2}$$

$$- 3\mu^{plagioclase}_{CaAl_2Si_2O_8} , \qquad (1.10)$$

and at equilibrium

$$\Delta G = 0.$$

If ΔG is less than zero then garnet, kyanite, and quartz
would form from anorthite. If it is greater than zero,
anorthite is stable. As in the previous case the subscripts
$Ca_3Al_2Si_3O_{12}$, SiO_2 etc. refer to components and the

superscripts kyanite, garnet etc. to <u>phases</u>, i.e. $\mu^{kyanite}_{Al_2SiO_5}$
refers to the chemical potential of the Al_2SiO_5 component in
the kyanite phase. Since μs are molar properties, it is
necessary to introduce the coefficients 3 for $CaAl_2Si_2O_8$ and
2 for Al_2SiO_5 because there are, respectively, 3 and 2
moles of these components involved in reaction (1.9).

In rocks, garnets are never pure $Ca_3Al_2Si_3O_{12}$, nor are
plagioclases ever pure $CaAl_2Si_2O_8$. In both cases they are
multicomponent solid-solutions of widely variable composition.
<u>However, it is very important to note that if there is
equilibrium between plagioclase solid-solution, garnet solid-
solution, kyanite (generally almost pure Al_2SiO_5) and quartz
(almost pure SiO_2) in any rock then ΔG for the reaction in eqn
(1.10) must equal zero.</u> This condition applies regardless
of the complexity of the phases present in the rocks. Although
the chemical potentials of the components are not the same in
the solid solutions as they are in the pure phases, reactions
and equilibrium conditions may be considered in terms of the
simple end-member components above (cf. reaction (1.9) and
the equilibrium condition (1.10)). The consideration of
equilibrium relations between the various simple components in
a multicomponent natural rock is an extremely powerful
technique when applied to geological problems, for it is
commonly possible to write several reactions in terms of
simple components for any given mineral assemblage. The more
reactions that can be written in terms of these simple
components, the greater the amount of information that can
be obtained from the mineral assemblage.

1.6 DETERMINATION OF μ AND G

In order to determine the equilibrium conditions for a given set of minerals, it is necessary to obtain numerical values of μ and G for the components and phases of interest. Although as stated above absolute values of these quantities cannot be determined, relative values can be obtained by choosing an arbitrary reference level. The Gibbs energy of a system or phase can be expressed as

$$G = H - TS. \qquad (1.11)$$

where H is the heat content or enthalpy of the system or phase and S is its entropy; T is the absolute temperature. The enthalpy, H, of the system is usually expressed relative to the enthalpies of the constituent elements. By convention the enthalpies of the constituent elements at 1 bar and the temperature of interest are taken to be zero. Thus a scale of relative enthalpies of compounds of these elements is established. Consider, for example, the enthalpy of pure forsterite Mg_2SiO_4. Taking the heat contents of Mg (crystal) Si (crystal) and O_2 (gas) to be zero at 1 bar and T, the enthalpy of Mg_2SiO_4 is obtained from the reaction

$$2Mg \ + \ Si \ + 2O_2 \ \rightleftharpoons Mg_2SiO_4$$
$$\text{crystal} \quad \text{crystal} \quad \text{gas} \quad \text{olivine}$$

$$\Delta H_{reaction} = H_{forst} - 2H_{Mg} - H_{Si} - 2H_{O_2}$$

$$= H_{forst} - 0 = -520.37 \text{ kcal } (-2177.23 \text{ kJ}) \text{ at 1 bar}$$
$$\text{and 298 K.}$$

The entropy of a system (or phase), S, may be regarded
as a measure of the degree of randomness or disorder in the
system. The greater the degree of disorder the higher the
value of the entropy. As an example let us consider a system
of large volume containing a small amount of a gaseous species.
The probability of finding a molecule of the gas within any
given volume of the system is very small because a few
molecules are dispersed throughout a large volume. The degree
of randomness of the system is therefore large and the molar
entropy of the gas is also large. As the volume of the system
is decreased the pressure of the gas increases, pressure
being inversely proportional to volume, and intermolecular
distances decrease.   By decreasing the total volume of the
system the probability of finding a molecule in any given
volume of gas is increased and entropy or randomness is
therefore decreased.   If pressure is increased sufficiently,
the gas may liquefy or even solidify. Using this argument,
it is apparent that molecules in the liquid state with low
volume and high density should have lower molar entropy
than those in the gaseous state, and that the molar entropy
in the solid state will, in general, be lower still. In
addition to the increased molecular density in liquid and
solid states relative to the gaseous state, the translational
and rotational degrees of freedom of molecules (also related
to randomness) are much lower in the liquid and solid states
than in the gaseous state because of the existence of
stronger intermolecular forces.

As a rule of thumb, the following relative entropy
scale may therefore be constructed :

$$(S_{gas})_{low\ P} > (S_{gas})_{high\ P} > S_{liquid} > S_{solid}. \tag{1.13}$$

Since G is an extensive property, both H and S are extensive properties. However, because it is intended to use molar values of G throughout the rest of this book, H_a and S_a will henceforward be taken to refer to the enthalpy per mole and entropy per mole of phase a. The units of H_a are therefore cal mol^{-1} (J mol^{-1}) and of S_a are cal mol^{-1} deg^{-1}, known otherwise as entropy units (e.u.). The analogous expression for the chemical potential of component i in phase a is

$$\mu_i^a = \bar{H}_i^a - T \bar{S}_i^a. \tag{1.14}$$

where $\bar{H}_i$ is the _partial_ molar enthalpy of component i in phase a and $\bar{S}_i$ is its partial molar entropy. The _partial_ molar enthalpy of component i is the enthalpy per mole of this component _in_ the phase of composition a. If phase a consists only of i then, as with G and μ, the molar enthalpy of a is equal to the partial molar enthalpy of i in a. If phase a is a mixture of components, this equality does not apply (see Chapter 3). Consider the reaction of forsterite with quartz to form enstatite (all phases pure) :

$$\underset{\text{forsterite}}{Mg_2SiO_4} + \underset{\text{quartz}}{SiO_2} \rightleftharpoons \underset{\text{enstatite}}{Mg_2Si_2O_6}. \tag{1.15}$$

The change in enthalpy of the system produced by reacting 1 mole of forsterite with 1 mole of quartz is given by

$$\Delta H = H_{enstatite} - H_{quartz} - H_{forsterite}. \tag{1.16}$$

For example, at 900 K, 1 bar,

$$\Delta H = -738\ 988 \quad + \quad 216\ 401 \quad + \quad 519\ 207 \quad = \quad -3380\ cal.$$
$$(3\ 091\ 930) \quad (905\ 422) \quad (2\ 172\ 360) \quad (-14\ 142\ J)$$

ΔH is the heat evolved as the reaction takes place at constant pressure.   Similarly the entropy change for reaction is

$$\Delta S = S_{enst} \ - \ S_{qz} \ - \ S_{fo}.$$

At 900 K, 1 bar,

$$\Delta S = \quad 86.12 \quad - \quad 26.09 \quad - \quad 61.88$$
$$(360.33) \quad (109.16) \quad (258.91)$$

$$= \ -1.85\ cal\ mol^{-1}\ deg^{-1}.$$
$$(-7.74\ JK^{-1}\ mol^{-1})$$

In order to determine whether or not reaction (1.15) should proceed from left to right or vice versa under a given set of P, T conditions, it is necessary to know ΔG under these conditions. This is given by

$$\Delta G = \mu^{enst}_{Mg_2Si_2O_6} \ - \ \mu^{fo}_{Mg_2SiO_4} \ - \ \mu^{qz}_{SiO_2}$$

or, since all phases are pure :

$$\Delta G = G_{enst} - G_{fo} - G_{qz}$$
$$= \Delta H - T\Delta S. \tag{1.17}$$

By analogy with the case of calcite and aragonite (section 3) :

$\Delta G < 0$ reaction proceeds to the right
$\Delta G > 0$ reaction proceeds to the left
$\Delta G = 0$ equilibrium.

In this case at 900 K, 1 bar we have

$$\Delta G = - 3380 + 900 \times 1.85 = - 1715 \text{ cal } (- 7176J).$$

So at 1 bar and 900 K, enstatite is produced from forsterite and quartz.

Hess's law

Since H, S (and volume V) are functions of state of the system, they may be measured indirectly in cases where there are experimental difficulties in their direct determination. For example, it is extremely difficult to produce diamond from graphite via the reaction

$$\underset{\text{graphite}}{C} \rightleftarrows \underset{\text{diamond}}{C} \qquad\qquad \Delta H_1 . \tag{1.18}$$

because of the high pressures and temperatures necessary to produce a measurable amount of reaction. The experimental difficulties involved in determining ΔH_1 directly are therefore enormous. However, the heats of combustion in oxygen of each polymorph can readily be determined.

$$\underset{\text{graphite}}{C} + \underset{\text{gas}}{O_2} \rightleftharpoons \underset{\text{gas}}{CO_2} \qquad \Delta H_2 \qquad\qquad (1.19)$$

$$\underset{\text{diamond}}{C} + \underset{\text{gas}}{O_2} \rightleftharpoons \underset{\text{gas}}{CO_2} \qquad \Delta H_3 \qquad\qquad (1.20)$$

Reversing reaction (1.20) gives

$$CO_2 \rightleftharpoons \underset{\text{diamond}}{C} + O_2 \qquad\qquad \Delta H = -\Delta H_3. \qquad (1.21)$$

Adding (1.19) and (1.21) gives the net reaction :

$$\underset{\text{graphite}}{C} + CO_2 + O_2 \rightleftharpoons \underset{\text{diamond}}{C} + CO_2 + O_2$$

$$\Delta H = \Delta H_2 - \Delta H_3$$

This is the same as

$$\underset{\text{graphite}}{C} \rightleftharpoons \underset{\text{diamond}}{C} \qquad \Delta H_1 = \Delta H_2 - \Delta H_3.$$

ΔH_1 may therefore be directly calculated from measured values of ΔH_2 and ΔH_3. All of the extensive variables G, S, H, and V are additive and all of them may be determined in analogous indirect ways.

1.7 TEMPERATURE DEPENDENCE OF H AND S, μ, AND G

If a substance is heated at constant pressure, then its heat content or enthalpy increases in approximate propor- tion to its increase in temperature. The relationship between temperature increase and enthalpy increase is different for different substances and is expressed by a characteristic constant of proportionality, the heat capacity C_p. If the substance is heated through a temperature interval dT then the increase in enthalpy, dH is given by

$$dH = C_p \, dT. \tag{1.22}$$

Since the enthalpies of reactants and products change with temperature, the enthalpy change of a reaction is also temperature dependent :

$$dH_A = (C_p)_A \, dT; \quad dH_B = (C_p)_B \, dT$$

$$d\Delta H = (C_p)_B \, dT - (C_p)_A \, dT = \Delta C_p \, dT \tag{1.23}$$

where $\Delta H = \Sigma H_{products} - \Sigma H_{reactants}$

$\Delta C_p = \Sigma C_p^{products} - \Sigma C_p^{reactants}$.

In tables of thermodynamic data (e.g. Robie and Waldbaum 1968) it is common to give enthalpies of minerals and gases at a reference temperature, usually 298.15 K (25.0°C), and in order to obtain values at higher tempera- tures it is necessary to integrate eqn (1.22) for each mineral.

To obtain H at any temperature T given H_{298} and C_p :

$$\int_{298}^{T} dH = \int_{298}^{T} C_p \, dT \tag{1.24}$$

$$H_T - H_{298} = \int_{298}^{T} C_p \, dT \tag{1.25}$$

If the heat capacity is independent of temperature, we have

$$H_T = H_{298} + C_p \, (T - 298). \tag{1.26}$$

In general, however, heat capacities are not independent of temperature and the integration is slightly more
complex. Heat capacities have been measured for most phases
of geological interest and are tabulated by Kelley (1960).
In cases where heat capacities are not known they can
generally be estimated with a reasonable degree of confidence
(section 1.12).

The general form of heat capacity at constant pressure
as a function of temperature is illustrated in Fig. 1.3.

As can be seen, C_p passes through zero at absolute zero
and approaches a constant value at high temperature. Within
the temperature range above 298.15 K, heat capacities can
be represented by the following simple equation :

$$C_p = a + bT + c/T^2 \tag{1.27}$$

where a, b, and c are experimentally determined constants
for the phase of interest and T is the absolute temperature.

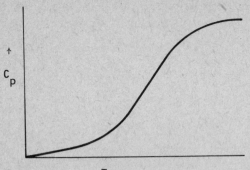

FIG. 1.3. General form of the heat capacity at constant pressure as a function of temperature.

To determine the heat content of a phase at any temperature T, substitution of (1.27) into (1.25) yields

$$H_T = H_{298} + \int_{298}^{T} (a + bT + c/T^2) \, dT. \qquad (1.28)$$

Integrating,

$$H_T = H_{298} + \left[aT + \frac{bT^2}{2} - \frac{c}{T} \right]_{298}^{T}. \qquad (1.29)$$

The temperature of interest, T, is substituted into the expression in the square brackets and the result added to H_{298}. A similar substitution is made for 298 K and the result subtracted from H_{298}. The total gives H_T.

<u>Example 1.</u>   To determine the enthalpy of low albite at 900 K
and 1 bar given the value at 298.15 K and 1 bar and the heat-
capacity function (in cal $deg^{-1} mol^{-1}$).

For albite,

$$C_p = 61.7 + 13.9 \times 10^{-3}T - \frac{15.01 \times 10^5}{T^2} \quad \text{(Kelley 1960)}.$$

$$H_{298} = -937\ 146\ cal\ mol^{-1} \quad \text{(Robie and Waldbaum 1968)}.$$

$$H_{900} = H_{298} + \left[aT + \frac{bT^2}{2} - \frac{c}{T} \right]_{298}^{900} \tag{1.30}$$

$$= -937\ 146 + \left[61.7T + \frac{13.9 \times 10^{-3}T^2}{2} + \frac{15.01 \times 10^5}{T} \right]_{298}^{900}.$$

Substitution of the values of T into expression (1.30)
above yields

$$H_{900} = -937\ 146 + 61.7 + \frac{13.9 \times 10^{-3}}{2} (900^2 - 298^2)$$

$$+ 15.01 \times 10^5 \left(\frac{1}{900} - \frac{1}{298} \right) = -898\ 360\ cal\ mol^{-1}.$$

Increasing the heat content of a phase also leads to
an increase in its entropy. If an amount of heat dH is
supplied at constant pressure we have :

$$dS = \frac{dH}{T}, \tag{1.31}$$

or, from (1.22)

$$dS = \frac{C_p \, dT}{T} \, . \tag{1.32}$$

So to obtain the entropy of a phase at T, given the value at 298.15 K, eqn (1.32) can be integrated as follows :

$$S_T - S_{298} = \int_{298}^{T} \left(\frac{a}{T} + b + \frac{c}{T^3} \right) dT \tag{1.33}$$

$$S_T = S_{298} + \left[a \ln T + bT - \frac{c}{2T^2} \right]_{298}^{T} \tag{1.34}$$

and, as with the expression for H_T, temperature T and temperature 298 K are substituted into the term in square brackets in order to obtain S_T.

Example 2. To calculate ΔH_{800} and ΔS_{800} for the reaction

$$
\begin{array}{ccc}
NaAlSi_3O_8 & \rightleftharpoons & NaAlSi_2O_6 + SiO_2 \\
\text{albite} & & \text{jadeite} \quad\quad \text{quartz}
\end{array} \tag{1.35}
$$

given the following data :

Phase	H_{298} cal mol^{-1}	S_{298} e.u.	C_p cal mol^{-1} deg^{-1}
Low Albite	- 937 146	50.20	$61.7 + 13.9 \times 10^{-3}T$ $- \dfrac{15.01 \times 10^5}{T^2}$
Jadeite	- 719 871	31.90	$48.16 + 11.42 \times 10^{-3}T$ $- \dfrac{11.87 \times 10^5}{T^2}$
Quartz	- 217 650	9.88	$11.22 + 8.2 \times 10^{-3}T$ $- \dfrac{2.7 \times 10^5}{T^2}$

$$\Delta H_{298} = + 937\ 146 - 719\ 871 - 217\ 650 = -375 \text{ cal}$$

$$\Delta S_{298} = -50.2 + 31.9 + 9.88 = -8.42 \text{ e.u.}$$

$$\Delta C_p = \Delta a + \Delta bT + \frac{\Delta c}{T^2}$$

$$= -2.32 + 5.72 \times 10^{-3}T + \frac{0.44 \times 10^5}{T^2} \text{ cal mol}^{-1} \text{ deg}^{-1}.$$

$$\Delta H_{800} = \Delta H_{298} + \int_{298}^{800} \Delta C_p \, dT$$

$$= -375 + \left[-2.32T + \frac{5.72}{2} \times 10^{-3} T^2 - \frac{0.44 \times 10^5}{T} \right]_{298}^{800}.$$

Substitution for T = 800 and T = 298 yields

$$\Delta H_{800} = - 375 - 80.6 + 581.55 = \underline{+ 125.95 \text{ cal}}.$$

Similarly,

$$\Delta S_{800} = \Delta S_{298} + \int_{298}^{800} \frac{\Delta C_p}{T} dT \qquad (1.37.$$

$$= - 8.42 + \left[- 2.32 \ln T + 5.72 \times 10^{-3} T - \frac{0.44 \times 10^5}{2T^2} \right]_{298}^{800}$$

and substituting for T as before

$$\Delta S_{800} = - 8.42 - 10.96 + 11.77 = - 7.61 \text{ e.u.}$$

From these results it can easily be seen that the assumption of $\Delta C_p = 0$ introduces extremely small errors in ΔG for the reaction involving pure jadeite, albite, and quartz :

$$\Delta G = \Delta H - T\Delta S.$$

Assuming $\Delta C_p = 0$ and using the 298 K values of ΔH and ΔS gives

$$\Delta G_{1 \text{ bar},800} = - 375 + 800 \times 8.42 = \underline{+ 6361 \text{ cal}}.$$

If the temperature dependence from the difference in heat

capacities is considered, the calculated value of $\Delta G_{1\ bar,\ 800}$
is 6214 cal (126 + 800 x 7.61). The error introduced by using
the 298 K values at 800 K is therefore less than 150 cal, which
is an order of magnitude less than the uncertainty in ΔH.

It is reasonable therefore to make the following
assumption about the value of $\Delta G_{1\ bar,T}$:

$$\Delta G_{1\ bar,T} = \Delta H_{1\ bar} - T\Delta S. \tag{1.38}$$

where ΔH and ΔS are effectively constant over a wide range of
temperature. This assumption is applicable virtually without
exception to <u>solid-solid</u> reactions over the range of temperatures of geological interest. Therefore for reactions
involving only solid phases, it is sufficient to know $\Delta H_{1\ bar}$
and ΔS at some temperature close to the temperature range
of interest, or even at 298 K in many cases, in order to be
able to calculate $\Delta G_{1\ bar,T}$ with sufficient accuracy. The
validity of this approximation obviates the necessity of
performing the heat-capacity integrations for reactions
involving only solids. The assumption that ΔC_p is zero is
also adequate for reactions involving gases and liquids, but
generally only over a relatively small temperature range
(200 - 300°C or so). It may or may not, depending on the
accuracy desired, be necessary to perform the heat-capacity
integrations in these cases.

1.8 PRESSURE DEPENDENCE OF μ AND G

At constant temperature T, the effect of pressure on
the free energy of phase a (G_a) is given by

$$\left(\frac{\partial G_a}{\partial P}\right)_T = V_a. \tag{1.39}$$

Similarly the change in chemical potential of component i in phase a is equal to the _partial_ molar volume of component i in phase a :

$$\left(\frac{\partial \mu_i^a}{\partial P}\right)_T = \bar{V}_i. \tag{1.40}$$

For a reaction involving a volume change ΔV, where

$$\Delta V = \Sigma V_{products} - \Sigma V_{reactants}$$

the analogous expression is

$$\left(\frac{\partial \Delta G}{\partial P}\right)_T = \Delta V. \tag{1.41}$$

Integration of (1.41) enables us to calculate ΔG for a reaction at any pressure P and the temperature of interest T, provided that ΔG at one pressure and temperature T is known.

1.9 TEMPERATURE AND PRESSURE DEPENDENCE OF μ AND G

The Gibbs energy of a reaction at any temperature and pressure may be calculated by combining eqns (1.36) and (1.37) (temperature dependence of ΔG) with (1.41) above :

$$(\Delta G)_{P,T} = \Delta H_{1 \text{ bar, } 298} + \int_{298}^{T} \Delta C_p \, dT$$

$$\tag{1.42}$$

$$- T \left(\Delta S_{298} + \int_{298}^{T} \frac{\Delta C_p}{T} \, dT \right) + \int_{1}^{P} \Delta V \, dP.$$

Evaluation of eqn (1.42) enables the conditions under which the phases are in equilibrium to be determined.

An additional simplifying assumption which can almost always be made for solid phases in geological calculations is that the volumes of solids are independent of pressure and temperature. In that case we have

$$\int_{1}^{P} V \, dP = (P-1) \, V. \tag{1.43}$$

Or, for a reaction involving solids only :

$$\int_{1}^{P} \Delta V_{solids} \, dP = (P-1) \, \Delta V_{solids}. \tag{1.44}$$

Although generally unnecessary, thermal expansion (α) and compressibility (β) may be taken into account by expressing volumes of solids in terms of these parameters (data in Clark 1966) :

$$V = V_{1 \text{ bar, } 298} + \alpha(T - 298) + \beta(P-1). \tag{1.45}$$

1.10 CALCULATION OF A REACTION BOUNDARY

Consider the reaction

$$NaAlSi_3O_8 \rightleftarrows NaAlSi_2O_6 + SiO_2$$
$$\text{albite} \qquad \text{jadeite} \qquad \text{quartz}$$

(1.46)

At equilibrium we have :

$$\Delta G = \mu^{jd}_{NaAlSi_2O_6} + \mu^{qz}_{SiO_2} - \mu^{alb}_{NaAlSi_3O_8} = 0$$

which, if the phases are all pure, is equivalent to

$$\Delta G = G_{jd} + G_{qz} - G_{alb} = 0.$$

Thermodynamic data for reaction (1.46) (pure phases) taken from Robie and Waldbaum (1968) are as follows :

$$\Delta H_{1,\,298} = + 937\ 146 \quad - 719\ 871 \qquad - 217\ 650$$
$$(3\ 921\ 020) - (3\ 011\ 940) - (910\ 650)$$

$$= - 375\ cal$$
$$= - (1.569\ kJ)$$

$$\Delta S_{298} = - 8.42\ e.u.$$
$$(- 35.23\ JK^{-1}\ mol^{-1})$$

Note that the enthalpy values quoted in Robie and Waldbaum are strictly for 1 atmosphere which is 1.013 bar. The error in ΔG introduced by assuming that the values are applicable to 1 bar is only

$$(1.013 - 1) \times \Delta V = -0.005 \text{ cal.}$$

For equilibrium at 298 K and pressure P we have the condition

$$\Delta G_{P,T} = \Delta H_{298} - 298 \Delta S_{298} + \int_1^P \Delta V \, dP = 0. \tag{1.47}$$

At a point on the equilibrium boundary which is at temperature dT above 298 K and pressure dP above P we have from (1.42)

$$\Delta G_{P + dP, 298 + dT} = \Delta H_{298} + \Delta C_p \, dT - (298 + dT) \left(\Delta S \right.$$

$$\left. + \frac{\Delta C_p \, dT}{298 + dT} \right) \tag{1.48}$$

$$+ \int_1^{(P + dP)} \Delta V \, dP = 0 \text{ (equilibrium).}$$

Equating (1.48) and (1.47),

$$- \Delta S \, dT + \Delta V \, dP = 0$$

$$\frac{dP}{dT} = \frac{\Delta S}{\Delta V}. \tag{1.49}$$

The Clausius-Clapeyron eqn (1.49) enables the slope of the

reaction boundary at any pressure and temperature of interest to be calculated from ΔS and ΔV.

If we make the assumption that ΔC_p is equal to 0 ($\Delta H_{1\ bar}$, ΔS constant) the equilibrium temperature at 1 bar is given by

$$\Delta G = 0 = \Delta H_{1\ bar} - T\Delta S$$

$$\Delta H_{1\ bar} = T\Delta S$$

$$T = \frac{-\ 375}{-\ 8.42} = \underline{44.5\ K.}$$

The equilibrium boundary could now be calculated at some higher pressure, P_1, using

$$\Delta G = 0 = \Delta H_{1\ bar} - T\Delta S + (P_1 - 1)\ \Delta V.$$

Alternatively, the 1 bar result may be extrapolated to higher pressure using the Clausius-Clapeyron equation and making the assumption that both ΔS and ΔV are constant at the 1 bar, 298 K values. This is generally reasonable for solid-solid reactions.

Before calculating dP/dT from the Clausius-Clapeyron equation, it is necessary to make the units of ΔV consistent with those of ΔS. If ΔS is in entropy units, the consistent units of ΔV are calories per bar (cal bar^{-1}) and the conversion factor from cm^3 is 1/41.84. (Note that this correction is not necessary if SI units are used since the units in this system are self-consistent.)   Thus

$$\Delta V \text{ cal bar}^{-1} = \Delta V \text{ cm}^3/41.84.$$

In this case

$$\Delta V = \frac{-16.98}{41.84} = -0.4058 \text{ cal bar}^{-1} = -16.98 \times 10^{-6} \text{m}^3$$

$$\frac{dP}{dT} = \frac{-8.42}{-0.4058} = \underline{+20.75 \text{ bars per degree}} \text{ (bar deg}^{-1})$$

$$= \left(\frac{-35.23 \text{ N m K}^{-1}}{-16.98 \times 10^{-6} \text{m}^3} \right.$$

$$\left. = 20.75 \times 10^5 \text{ N m}^{-2} \text{ K}^{-1} = 20.75 \text{ bar K}^{-1} \right)$$

At 298 K, therefore, P = 1 + 20.75 × (298 - 44.5)

$$= \underline{5260 \text{ bar}}.$$

At 900 K, P = 1 + 20.75 × (900 - 44.5)

$$= \underline{17750 \text{ bar}}.$$

$$= \underline{17.75 \text{ kbar}}.$$

The phase diagram calculated from these 298 K data is shown in Fig. (1.4) . This figure indicates that albite is stable up to 17.75 kbar at 900 K and 5.26 kbar at 298 K.   If we were to use the 900 K values of enthalpy and entropy (tabulated in Robie and Waldbaum) to calculate the phase diagram, then the calculated pressure at 900 K would be 17.05 kbar and at 298 K, 6.54 kbar.   It is apparent from these figures that the assumption of ΔCp equal to zero is quite reasonable over large temperature

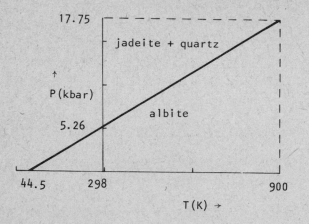

FIG. 1.4. Equilibrium boundary between albite, jadeite, and quartz calculated from 1 bar, 298 K values (all assumed constant) of enthalpy, entropy, and volume.

ranges for this solid-solid reaction. The error introduced at 900 K by using the 298 K values is within the experimental error in the solid-media apparatus which is used experimentally to determine phase diagrams at high pressure.

1.11 MORE ABOUT ENTROPY

The total entropy or disorder of a system or phase may be considered, qualitatively, as having two contributions :

(a) Thermal disorder, due to motion, rotation, and vibration of atoms and molecules in solids, liquids and gases.

(b) <u>Entropy of mixing</u>, due to mixing more than one
type of atom or molecule in a liquid or gas or by mixing
more than one type of atom on any one type of site in a
solid. As an example, let us consider olivine solid
solutions between forsterite Mg_2SiO_4 and fayalite Fe_2SiO_4.
In the olivine structure there are two types of cation
position, M1 and M2, occupied by Mg^{2+} in forsterite and
Fe^{2+} in fayalite.   In intermediate members of the series,
$(Mg_xFe_{1-x})_2 SiO_4$, magnesium and iron atoms are randomly
mixed on these two positions. If we were to pick any M1 or
M2 position, the probability of it being occupied by a
magnesium atom is x and the probability of it being occupied
by an iron atom is $(1-x)$.   Mixing of the iron and magnesium
atoms therefore introduces substantial disorder or randomness
which is not present in the pure end-members. In the end-
members all M1 and M2 positions are occupied by Mg
(forsterite) or Fe (fayalite).   The entropy-of-mixing
contributions to the entropies of phases are obviously very
important in rocks because almost all natural minerals are
multicomponent solid solutions.

The mixing properties of solid-solutions will be
discussed in more detail in Chapter 3. For the moment let
us return to the entropies of <u>pure</u> phases which do not have
entropies of mixing.

Entropies of pure phases.

Given the entropy of a mineral at 298 K and its heat
capacity, the entropy at any other temperature may be
calculated from

$$\int_{298}^{T} dS = \int_{298}^{T} \frac{C_p}{T} \, dT. \qquad (1.50)$$

An analogous integration gives the entropy of the mineral
relative to the value at the absolute zero of temperature
$(- 273.15\,^{\circ}C)$:

$$S_T - S_0 = \int_{0}^{T} \frac{C_p}{T} \, dT. \qquad (1.51)$$

At the absolute zero of temperature, however, all thermal
motion in a solid ceases and there can be no disorder due
to lattice vibrations or other atomic motions. This leads
to the third law of thermodynamics - the entropy of a pure
crystalline substance at absolute zero is zero - and gives
from eqn (1.51)

$$S_T = \int_{0}^{T} \frac{C_p}{T} \, dT. \qquad (1.52)$$

The entropy of a pure crystalline substance can therefore
be obtained directly from heat-capacity measurements by
assuming that S_0 (at 0 K) is zero. Robie and Waldbaum
(1968) tabulate 'third-law' entropy values for most
crystalline phases of geological interest at 298.15 K and
higher temperatures.

In cases where the 'third-law' entropy of a pure
mineral is not known, it is possible to make an entropy

estimate based on a summation of the values for the component oxides with an empirical correction for the difference
in volumes. There are a number of empirical equations which
have been used for this purpose and two which are applicable
at 298.15 K are :

$$S_{298.15} = (S_{ox})_{298.15} + 0.5 (V - V_{ox}) \text{ e.u.} \qquad (1.53)$$

$$S_{298.15} = (S_{ox})_{298.15}\left(\frac{1 + (V/V_{ox})}{2}\right) \text{ e.u.} \qquad (1.54)$$

In these equations V and $S_{298.15}$ refer to the volume (in cm^3)
and entropy of the silicate, S_{ox} and V_{ox} to the sums of the
298.15 K entropies and volumes (in cm^3) of the oxide
constituents of the silicates. Application of these equations
to several phases in the system $CaO-Al_2O_3-SiO_2$ yields the
estimates of the 298 K entropies which are shown in Table 1.1.
The latter are in reasonably good agreement with the experimentally determined values of $S_{298.15}$. The estimates can
often be improved by using different empirical coefficients
for the different types of silicates (one for garnets,
another for pyroxenes etc.) or by generating entropy values
for complex silicates using those for simple silicates of
similar structure, e.g. using pyroxene entropies to calculate
amphibole entropies with volume corrections as in eqns
(1.53) and (1.54).

An analogous expression to (1.54) can be used to
estimate the heat capacities of mineral phases for which
there are no data available. These estimates are, if

TABLE 1.1

Observed and calculated 298 K entropy values

for some phases in the system $CaO-Al_2O_3-SiO_2$

Phase	$S_{298}^{observed}$ e.u.	S_{298}^{ox} e.u.	$S_{298}(1.53)^*$	$S_{298}(1.54)^\dagger$
$CaAl_2Si_2O_8$ anorthite	48.45	41.44	47.48	44.26
$Ca_3Al_2Si_3O_{12}$ grossular	57.7	70.32	61.0	65.8
Ca_2SiO_4 olivine	28.8	28.88	30.33	29.62
Al_2SiO_5	20.02	22.06	19.97	21.11

*Entropy calculated using eqn (1.53)

†Entropy calculated using eqn (1.54)

anything, even more accurate than the entropy estimates discussed above. The heat capacity function is

$$C_p = a + bT + c/T^2.$$

Each of the constants a, b, and c in this equation is estimated in the following way :

$$a = a_{ox} \left(\frac{1 + (V/V_{ox})}{2} \right).$$ (1.55)

and so on. In conclusion it may be stated that entropies and heat capacities are available, or can be estimated with reasonable confidence, for most pure mineral phases of geological interest.

SUMMARY - CHAPTER 1

1.   For any reaction involving components of minerals, fluids or melts :

$$\begin{array}{cccc} wA + xB \gtrless yC + zD \\ \text{phase} \quad a \quad\quad b \quad\quad c \quad\quad d \end{array}$$ (1.56)

the condition of equilibrium at any pressure and temperature is :

$$y\mu_C^c + z\mu_D^d - w\mu_A^a - x\mu_B^b = \Delta G = 0$$

where μ_C^c refers to the chemical potential of component C in phase c etc., and w, x etc. to number of moles of components involved in the reaction. If the phases are pure then μs can be replaced by molar free energies (G_c etc.) of the phases at the pressure and temperature of interest.

2.   The chemical potentials or free energies can be calculated at any pressure and temperature given molar

(or partial molar) enthalpies H, entropies S, heat capacities C_p, and volumes V :

$$G_{P,T} = H_{1\ bar,298} + \int_{298}^{T} C_p\ dT - T \left(S_{298} + \int_{298}^{T} \frac{C_p}{T}\ dT \right)$$

$$+ \int_{1}^{P} V\ dP.$$

3.   The general form of the heat capacity equation is

$$C_p = a + bT + c/T^2$$

where a, b, and c are experimentally determined constants.

4.   For reactions involving solids only, it is generally valid to make the assumption that ΔC_p ($\Sigma (C_p)_{products} - \Sigma (C_p)_{reactants}$) equals 0 and that ΔV is a constant. Making these assumptions gives

$$(\Delta G)_{P,T} = \Delta H_{1\ bar} - T\Delta S + (P-1)\ \Delta V$$

where ΔH, ΔS, and ΔV are constants. For reaction (1.56) above, ΔH, ΔS etc. are given by

$$\Delta H = yH_c + zH_d - wH_a - xH_b.$$

etc.

5. The slope of a reaction boundary in P-T space is given
    (at any point) by the Clausius-Clapeyron equation

$$\frac{dP}{dT} = \frac{\Delta S}{\Delta V}$$

where ΔS and ΔV are entropy and volume changes of the
reaction at the pressure and temperature of interest.

6. The entropies of pure crystalline phases at any tempera-
    ture may be obtained from the third law of thermodynamics :

$$S_T = \int_0^T \frac{C_p}{T} \, dT.$$

7. Where entropies or heat capacities of silicates are not
    known, reasonable estimates may be made by summing the
    entropies of the constituent oxides and making volume
    corrections.

PROBLEMS - CHAPTER 1

1. Given values of S_{298} and V_{298} for the following minerals
    consider reactions (i) to (iv) and state whether products
    or reactants will be favoured by (a) increase of tempera-
    ture, or (b) increase of pressure. (Determine whether ΔG
    should become more positive or more negative with small
    increases in T and P.)

	S (J K^{-1} mol^{-1})	S e.u.	V cm^3
Albite (low)	(210.04)	50.20	100.27
Kyanite	(83.76)	20.02	44.09
Sillimanite	(96.11)	22.97	49.9
Quartz (alpha)	(41.34)	9.88	22.69
Silica glass	(47.44)	11.33	27.27
Jadeite	(133.5)	31.90	60.40
Anorthite	(202.7)	48.45	100.79
Wollastonite	(82.01)	19.60	39.93
Grossular	(241.4)	57.7	125.3

(i) $(Al_2SiO_5) \rightleftharpoons (Al_2SiO_5)$
      kyanite   sillimanite

(ii) $CaAl_2Si_2O_8 + 2CaSiO_3 \rightleftharpoons Ca_3Al_2Si_3O_{12} + SiO_2$
      anorthite   wollastonite   grossular   quartz

(iii) $(SiO_2) \rightleftharpoons (SiO_2)$
      quartz   glass

(iv) $NaAlSiO_8 \rightleftharpoons NaAlSi_2O_6 + SiO_2$
      albite   jadeite   quartz

2.   Assuming $\Delta C_p = 0$ and that ΔV is independent of
pressure and temperature, calculate and sketch the
Al_2SiO_5 phase diagram from the following data (derived
from the work of Holdaway (1971) - see Chapter 7).

kyanite $\rightleftharpoons$ sillimanite

$(\Delta H)_{1\ bar}$ = + 1770 cal (7406 J) ΔS = + 2.95 e.u.
$$(12.34\ J\ K^{-1})$$

ΔV = + 5.81 cm^3 = 0.1389 cal bar^{-1}

andalusite $\rightleftharpoons$ sillimanite

$(\Delta H)_{1\ bar}$ = + 730 cal (3054 J) ΔS = + 0.71 e.u.
$$(2.971\ J\ K^{-1})$$

ΔV = - 1.63 cm^3 = - 0.03896 cal bar^{-1}

Assume all ΔH, ΔS, ΔV are independent of P and T. Note that the third reaction andalusite $\rightleftharpoons$ kyanite must be calculated using the data given.

3.   Calculate S_{800} for albite ($NaAlSi_3O_8$) given :

S_{298} = 50.20 e.u. (Robie and Waldbaum 1968)
$$(210.04\ J\ K^{-1})$$

C_p = 61.7 + 13.9 x 10^{-3}T - 15.01 x $10^5/T^2$
$$\text{cal } deg^{-1}\ mol^{-1}\ \text{(Kelley 1960)}$$

= (258.2) + (58.2 x 10^{-3}T) - (62.80 x $10^5/T^2$)
$$J\ K^{-1}\ mol^{-1}$$

SOLUTIONS TO PROBLEMS - CHAPTER 1

1.   Estimate the effects of temperature and pressure on
ΔG of the reactions using

$$\Delta G = \Delta H_{1 \, bar,T} - T\Delta S + (P-1) \, \Delta V.$$

ΔG, ΔH, ΔS, ΔV all refer to $(\Sigma \mu_{prods} - \Sigma \mu_{reacts})$,

$(\Sigma H_{prods} - \Sigma H_{reacts})$ etc.

(i)   kyanite $\rightleftharpoons$ sillimanite

ΔS is positive, so ΔG becomes increasingly negative
with increasing T. Sillimanite is therefore
favoured with increasing T. ΔV is positive, so ΔG
becomes more positive with increasing P, and
reactants (kyanite) are favoured.

(ii) ΔV negative, ΔS negative. Increasing P
favours products; increasing T favours reactants.

(iii) ΔV positive, ΔS positive. Increasing P
favours reactants; increasing T favours products.

(iv) ΔV negative, ΔS negative. Increasing P
favours products; increasing T favours reactants.

2.   The Al_2SiO_5 phase diagram, calculated assuming $\Delta C_p = 0$
and ΔV constant.

(a)   kyanite $\rightleftharpoons$ sillimanite

$\Delta H_{1 \, bar}$ = 1770 cal (7406 J) ΔS = + 2.95 e.u.

$\qquad\qquad\qquad\qquad\qquad$ (+ 12.34 J K^{-1})

$\Delta V = 0.1389$ cal bar^{-1} (5.81 cm^3)

Equilibrium temperature at 1 bar $= \dfrac{\Delta H_{1 \text{ bar}}}{\Delta S} = 600$ K,

$\dfrac{dP}{dT} = \dfrac{\Delta S}{\Delta V} = \underline{21.238 \text{ bar deg}^{-1}}$

(b)   andalusite $\rightleftharpoons$ sillimanite

Equilibrium temperature at 1 bar = 1028 K,

$$\frac{dP}{dT} = \underline{-18.224 \text{ bar deg}^{-1}}$$

For the reaction

kyanite $\rightleftharpoons$ andalusite

ΔH, ΔS, ΔV may be obtained by subtracting the values for the andalusite $\rightleftharpoons$ sillimanite reaction from those for the kyanite $\rightleftharpoons$ sillimanite reaction :

$\Delta H = 1770 - 730 = 1040$ cal $= (4351$ J$)$

$\Delta S = 2.95 - 0.71 = 2.24$ e.u. $= (9.37$ J K$^{-1})$

$\Delta V = 0.1389 + 0.03896 = 0.17786$ cal bar^{-1} (7.44×10^{-6} m^3)

Equilibrium temperature at 1 bar = 464 K,

$$\frac{dP}{dT} = \underline{12.594 \text{ bar deg}^{-1}}.$$

The resulting phase diagram is shown overleaf.

Note that the low pressure part of the curve
kyanite ⇄ sillimanite is in a field where andalusite
is more stable than either of these phases. The (ky-sill)
curve is therefore metastable in this region, at pressures
below the invariant 3-phase point. The same applies to
the high temperature part of kyanite ⇄ andalusite (in
the sillimanite field) and the high pressure part of
andalusite ⇄ sillimanite (in kyanite field). In this
way the distribution of stable curves around the invariant
point may be deduced.

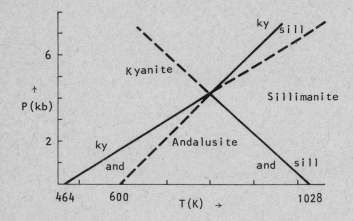

The position of the invariant point may be obtained
by writing the equations for any two curves in terms
of P and T and solving simultaneously, e.g. for ky-and,

$$T = 600 + (P-1)\left(\frac{1}{21.238}\right),$$

where $(1/21.238)$ is the slope in degrees per bar. For and - sill

$$T = 1028 - (P-1)\left(\frac{1}{18.224}\right).$$

Solving the two equations simultaneously yields

$$428 = (P-1)\left(\frac{1}{21.238} + \frac{1}{18.224}\right)$$

P = 4198 bar.

Substituting back in either of the other two equations gives

T = 798 K.

3.   To calculate S_{800} for albite,

$$C_p = 61.7 \quad + 13.9 \times 10^{-3}\ T \ - 15.01 \times 10^5/T^2$$
$$\text{cal deg}^{-1}\ \text{mol}^{-1}$$

$$(258.2) + (58.2 \times 10^{-3}\ T) - (62.80 \times 10^5/T^2$$
$$\text{J K}^{-1}\ \text{mol}^{-1})$$

$$\int_{298}^{800} dS = \int_{298}^{800} \frac{C_p}{T}\ dT$$

$$S_{800} - S_{298} = \int_{298}^{800}\left(\frac{61.7}{T} + 13.9 \times 10^{-3} - \frac{15.01 \times 10^5}{T^3}\right)\ dT$$

$$= \left[61.7 \ln T + 13.9 \times 10^{-3} T + \frac{15.01 \times 10^5}{2T^2} \right]_{298}^{800} .$$

Substituting T = 800 and evaluating the term in square brackets gives 424.73 e.u. Substituting T = 298 gives 364.1 e.u.   Therefore

$$S_{800} = 50.2 + 424.73 - 364.1 = \underline{110.83 \text{ e.u.}}$$
$$(463.71 \text{ J K}^{-1}).$$

2. Standard states, activities, and fugacities

2.1 INTRODUCTION

In Chapter 1 we introduced the concept of chemical potential, μ, as a measure of the tendency or otherwise of chemical species to react under certain conditions of pressure, temperature, and composition. The use of a reference state for the thermodynamic properties of components was also discussed, the reference point used being 298 K and 1 bar. Given values of $H_{1 \text{ bar},298}$, $S_{1 \text{ bar}, 298}$ and C_p for a pure phase, it is possible to calculate the chemical potential of the component of the pure phase at 1 bar and any temperature. If the volume of the phase is also known, μ can be calculated at any pressure and temperature.

When dealing with thermodynamic problems in geology it is generally convenient to take some reference value of the chemical potential μ_i of a component (at some pressure P_1, temperature T_1, and composition) and to refer values of μ_i in the calculations to this <u>standard state</u> value. The standard state chosen is solely a matter of convenience for any particular calculation; it has no effect on the

final result. Because of the availability of tabulated
thermodynamic data it is (as will become apparent) generally
most convenient to use standard states of pure components at
1 bar and the temperature of interest or at the pressure and
temperature of interest rather than 1 bar, 298 K. This
chapter is devoted to a discussion of some of the different
types of standard state and to the manipulation of μ_i using
various standard states.

2.2 DEFINITION OF ACTIVITY

The <u>activity</u> of component i in phase j (μ_i^j) reflects the
difference between the chemical potential of i in phase j
under given conditions of P,T and composition of interest, and
μ_i^0 in the standard state (at P^0, T^0). Algebraically the
definition is

$$\mu_i^j = \mu_i^0 + RT \ln a_i^j \qquad (2.1)$$

where a_i^j is the activity of component i in phase j and ln
refers to natural logarithms.

Rearranging (2.1), the expression for activity becomes

$$a_i^j = \exp\left(\frac{\mu_i^j - \mu_i^0}{RT}\right) \qquad (2.2)$$

For the moment eqns (2.1) and (2.2) will be accepted
as definitions of activity; the reasons for their form
will become apparent in succeeding sections.

We shall proceed by determining a_i^j and μ_i^j for three different types of standard state.

2.3 STANDARD STATE AT 1 BAR AND TEMPERATURE OF INTEREST

Consider a pure phase j which is composed solely of component i. At 1 bar and any temperature T, the chemical potential of i in the pure phase is given by :

$$(\mu_i^j)_{1,T} = (G_j)_{1,T} = (H_j)_{1,T} - T(S_j)_T \qquad (2.3)$$

At P bar and temperature T, μ_i^j in pure j may be obtained from

$$(\mu_i^j)_{P,T} = (\mu_i^j)_{1,T} + \int_1^P \bar{V}_i^j \, dP = (\mu_i^j)_{1,T} + \int_1^P V_j \, dP. \qquad (2.4)$$

(For a pure phase the partial molar volume of its constituent component is equal to the molar volume.)

Taking the standard state of component i to be pure i in phase j at 1 bar and temperature T, μ_i^0 (standard state chemical potential of i) is given by :

$$\mu_i^0 = (\mu_i^j)_{1,T}. \qquad (2.5)$$

Hence the chemical potential of i in pure j at P bar and T is

$$(\mu_i^j)_{P,T} = \mu_i^0 + \int_1^P V_j \, dP. \qquad (2.6)$$

Comparing eqn (2.6) with eqn (2.1) it is apparent that the
activity of component i in phase j at P bar and T may be
obtained from

$$RT \ln a_i^j = \int_1^P V_j \, dP. \tag{2.7}$$

If the phase under consideration is a solid, its volume may
be considered, approximately, to be independent of pressure.
In this case integration of the right-hand side of eqn (2.7)
gives

$$RT \ln a_i^j = V_j \, (P-1) \tag{2.8}$$

If phase j is a gas instead of a relatively incompressible
solid, the assumption that volume is independent of pressure
only holds over very small pressure ranges. So, for a
gaseous species, V_j has to be expressed as a function of
pressure before integrating eqn (2.7) to obtain activity at
Pbars and T. If phase j behaves as a perfect gas, the
relationship between volume and pressure is given by the
familiar equation of state :

$$PV_j = RT \tag{2.9}$$

or

$$V_j = \frac{RT}{P}. \tag{2.10}$$

Substituting (2.10) into (2.7) produces the relationship

$$RT \ln a_i^j = \int_1^P \frac{RT}{P} \, dP \qquad (2.11)$$

which can now be integrated to give

$$RT \ln a_i^j = RT \ln P - RT \ln 1 \qquad (2.12)$$

$$RT \ln a_i^j = RT \ln P/1 \qquad (2.13)$$

So the activity of a gaseous species in a pure perfect gas is, taking the standard state to be the pure gas at 1 bar and T, equal to the pressure :

$$a_i^j = P/1 \qquad (2.14)$$

In the more general case of a perfect gas with standard state pressure (P^0) not equal to 1 bar, the analogous expression is :

$$a_i^j = P/P^0. \qquad (2.15)$$

The reason for the form of the definition of activity (eqn (2.1)) is now apparent from eqns (2.12) and (2.13). From (2.15) it can be seen that the activity of a perfect gas component (relative to a 1 bar/T standard state) is the ratio of its pressure to that of the component in its standard state. More generally, the activity of any component is the ratio of the fugacity of the component to that in its standard state :

$$a_i^j = f_i / f_i^0$$

Fugacities have the dimensions of pressure and contain corrections for non-ideal behaviour (see section 2.8).

2.4 STANDARD STATE AT P BAR AND T

Consider the example discussed above, a phase j which is composed solely of component i. At P bar and T, the chemical potential of i in pure j may be obtained from (2.3) and (2.4) :

$$(\mu_i^j)_{P,T} = (H_j)_{1,T} - T(S_j)_T + \int_1^P V_j \, dP. \qquad (2.16)$$

If the standard state of component i is taken to be pure j at P bar and T, then μ_i^0 is obviously equal to $(\mu_i^j)_{P,T}$:

$$\mu_i^0 = (\mu_i^j)_{P,T} \qquad (2.17)$$

Comparing (2.1) and (2.17), the activity of i in j at P bar and T(K) is given by

$$RT \ln a_i^j = 0$$
$$(2.18)$$
$$a_i^j = 1 \text{ (since } \ln 1 = 0).$$

Thus the activity of the component of a pure phase at P bar and T(K), taking the standard state to be the pure phase at P and T is 1. The activity of any component in its standard state is 1.

2.5 STANDARD STATE AT 1 BAR AND 298 K

If the standard state of component i is taken to be pure j at 1 bar and 298 K, then μ_i^0 may be obtained from

1 bar, 298 K values of the enthalpy and entropy of j :

$$\mu_i^0 = (G_j)_{1,298} = (H_j)_{1,298} - 298 \ (S_j)_{298}. \tag{2.19}$$

The chemical potential of i in pure j at P bar and T must, as before, be given by

$$(\mu_i^j)_{P,T} = (H_j)_{1,T} - T \ (S_j)_T + \int_1^P V_j \ dP. \tag{2.20}$$

In order to determine the activity of i in j using this 1 bar, 298 K standard state, it is necessary to express $(H_j)_{1,T}$ and $(S_j)_T$ in terms of $(H_j)_{1,298}$, $(S_j)_{298}$ and C_p :

$$(\mu_i^j)_{P,T} = (H_j)_{1,298} + \int_{298}^T (C_p)^j \ dT - T \left((S_j)_{298} + \int_{298}^T \frac{(C_p)^j}{T} \ dT \right) + \int_1^P V_j \ dP. \tag{2.21}$$

Comparing (2.19) and (2.21) with (2.1) gives the activity of i in j for this standard state :

$$RT \ln a_i^j = \int_{298}^T (C_p)^j \ dT - (T - 298) \ (S_j)_{298} - T \int_{298}^T \frac{(C_p)^j}{T} \ dT + \int_1^P V_j \ dP. \tag{2.22}$$

Note that in each of the three cases, $(\mu_i^j)_{P,T}$ is the same and that the actual value of chemical potential is dependent solely on the pressure, temperature, and composition of the system; it has a unique value and is independent of the standard state chosen. The activity of component i in phase j depends greatly, however, on the standard state which is used.

The choice of standard state is solely a matter of convenience for ease of calculation and does not affect the results. In the last case discussed, in order to calculate activity and $(\mu_i^j)_{P, T}$, it is necessary to use $(H_j)_{1,298}$ and $(S_j)_{298}$ and to integrate the heat capacity function $(C_p)^j$. If a 1 bar, T or P, T standard state is used, it is not necessary to integrate the heat-capacity function in order to estimate a_i^j and $(\mu_i^j)_{P,T}$ provided $(\mu_i^j)_{1,T}$ is known. Since the tables of Robie and Waldbaum (1968) give values of S_T and (H_T-H_{298}) at 100° intervals, H_T, S_T, and (μ_i^j) may in many cases be calculated without the necessity of integrating C_p. In these cases it is easier in performing calculations to use 1 bar, T or P bar, T standard states for components and to take other data from the Robie and Waldbaum tables.

2.6 CALCULATION OF A REACTION BOUNDARY INVOLVING A FLUID PHASE

At high temperatures in metamorphosed siliceous lime-stones, calcite breaks down by reaction with quartz to form wollastonite and carbon dioxide :

$$CaCO_3 + SiO_2 \rightleftharpoons CaSiO_3 + CO_2 \qquad (2.23)$$
calcite quartz   wollastonite   fluid

The coexistence of calcite, quartz, wollastonite and fluid in equilibrium necessitates the condition :

$$\mu^{woll}_{CaSiO_3} + \mu^{fl}_{CO_2} - \mu^{cc}_{CaCO_3} - \mu^{qz}_{SiO_2} = \Delta G = 0 \qquad (2.24)$$

Substituting eqn (2.1) into eqn (2.24) for each of the four components, $CaSiO_3$, CO_2, $CaCO_3$ and SiO_2 gives

$$\mu^0_{CaSiO_3} + RT \ln a^{woll}_{CaSiO_3} + \mu^0_{CO_2} + RT \ln a^{fl}_{CO_2} - $$
$$- \mu^0_{CaCO_3} - RT \ln a^{cc}_{CaCO_3} - \mu^0_{SiO_2} - RT \ln a^{qz}_{SiO_2} = 0. \qquad (2.25)$$

Rearranging eqn (2.25) and collecting logarithmic terms on the right-hand side results in an equation known as the Van't Hoff isotherm :

$$\mu^0_{CaSiO_3} + \mu^0_{CO_2} - \mu^0_{CaCO_3} - \mu^0_{SiO_2} = \Delta G^0$$

$$= - RT \ln \left(\frac{a^{woll}_{CaSiO_3} \cdot a^{fl}_{CO_2}}{a^{cc}_{CaCO_3} \cdot a^{qz}_{SiO_2}} \right). \qquad (2.26)$$

Eqn (2.26) may be used to calculate the pressure-temperature conditions of equilibrium between calcite, quartz, wollastonite, and CO_2 for either pure one-component phases, or for 'natural' multicomponent phase compositions. Let us consider the application of eqn (2.26) to this equilibrium using two different sets of standard states.

(a)   Standard states at 1 bar and T

Consider the coexistence of pure calcite, quartz, wollastonite, and CO_2 with standard states of all components as the pure phase at 1 bar and T(K). For this standard state ΔG^0 is given by

$$\Delta G^0_{1,T} = \Delta H^0_{1,T} - T\Delta S^0_T \qquad (2.27)$$

where $\Delta H^0_{1,T}$ and ΔS^0_T are the differences between enthalpies and entropies of pure reactants and pure products at 1 bar and T.

   For a preliminary calculation we shall use the values of $\Delta H^0_{1,T}$ and ΔS^0_T given by Robie and Waldbaum (1968) at 900 K and assume $\Delta C_p = 0$.

$$\Delta H^0_{1,900} = (H^{woll}_{900} + H^{CO_2}_{900} - H^{cc}_{900} - H^{qz}_{900})$$

$$= - 389\ 591 - 94\ 267 + 286\ 800 + 216\ 401$$

$$= 19\ 343\ \text{cal}\ (80\ 931\ \text{J})$$

$$\Delta S^0_{900} = (S^{woll}_{900} + S^{CO_2}_{900} - S^{cc}_{900} - S^{qz}_{900}) = 35.07\ \text{e.u.}$$

$$(146.7\ \text{J K}^{-1}\ \text{mol}^{-1}).$$

Taking the standard states to be at 1 bar and T in each case, the activities of solid components to be substituted into eqn (2.25) may be obtained from (2.8) (assuming V_i constant) :

$$RT \ln a_{CaCO_3} = (P-1) V_{cc}$$

$$RT \ln a_{CaSiO_3} = (P-1) V_{woll} \text{ etc.}$$

(2.28)

Assuming that CO_2 behaves as a perfect gas and therefore that the activity of CO_2 is equal to pressure, eqn (2.26) becomes

$$\Delta H^0_{1,T} - T\Delta S^0_T = - RT \ln P_{CO_2} - (P-1) V_{woll} +$$

$$+ (P-1) V_{cc} + (P-1) V_{qz}.$$

(2.29)

Defining ΔV_{solids} as follows :

$$\Delta V_{solids} = V_{woll} - V_{cc} - V_{qz} = - 19.69 \text{ cm}^3$$

$$= - 0.47065 \text{ cal bar}^{-1} \text{ (Robie and}$$

Waldbaum)

(2.30)

and substituting the 1 bar, 900 values of enthalpy and entropy above into eqn (2.29) gives

$$19\ 343 - T (35.07) = - RT \ln P + (P-1)\ 0.47065.$$

(2.31)

The easiest way to calculate the equilibrium boundary is to substitute some particular value of pressure P into eqn (2.31) and solve for T. For example, at 5000 bar ($R = 1.987$ cal deg^{-1} mol^{-1})

$$19\ 343 - T (35.07) = - RT \ln 5000 + 4999 (0.47065)$$

$$19 \ 343 - T \ (35.07) = - \ 1.987 \ T \ 8.5172 + 2353$$
$$T = \underline{936 \ K}$$

At 2000 bar T = 922 K; at 1000 bar T = 884 K; at 1 bar, T = 552 K.

The above values have all been calculated by assuming that $\Delta C_p = 0$ <u>in the temperature range of interest</u>. This assumption may readily be checked by repeating the calculations using the actual thermodynamic data for 1000 K (for P = 5000 bar and 2000 bar), 800 K (P = 1000 bar) and 500 and 600 K (P = 1 bar). In the cases of the higher pressure points where the enthalpy and entropy data are only being extrapolated over 20-40°, the agreement between the 900 K and the 1000 or 800 K data is within 1°. The extrapolation to 550 K produces a larger error, however, the results obtained by using $\Delta H^o_{1,500}$, ΔS^o_{500}, and $\Delta H^o_{1,600}$, ΔS^o_{600} being 562 K i.e. 10° higher than that calculated from the 900 K data. It is apparent from these results that the assumption $\Delta C_p = 0$ when used over a temperature range of 200 - 300°C does not produce large errors in calculated equilibrium conditions.

Adopting an equilibrium temperature of 562 K at 1 bar and using the equilibrium temperatures at other pressures calculated from $\Delta H^o_{1,900}$ and ΔS^o_{900} gives the reaction boundary shown in Fig. 2.1. It should be noted that, in marked contrast to solid-solid reactions, the slopes of the boundaries of reactions involving fluid components are strongly dependent on pressure. From Chapter 1, the slope of a P-T curve is given by the Clausius-Clapeyron equation

$$\frac{dP}{dT} = \frac{\Delta S}{\Delta V} \cdot \qquad (2.32)$$

In this case ΔS and ΔV at any point on the curve are obtained from the values of S_i and V_i at the pressure and temperature of interest :

$$\Delta S = S_{woll} + S_{CO_2} - S_{cc} - S_{qz}$$

$$\Delta V = V_{woll} + V_{CO_2} - V_{cc} - V_{qz}.$$

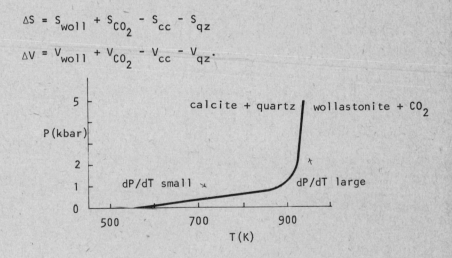

FIG. 2.1. Calculated P-T conditions of the reaction calcite + quartz $\rightleftharpoons$ wollastonite + CO_2. Note the influence of pressure on the slope of the equilibrium boundary.

Since the fluid component CO_2 is very much more compressible than the solid components, ΔV decreases rapidly with increasing pressure and this decrease produces an increase in dP/dT. (ΔS also decreases to some extent with increasing pressure, but not as rapidly as ΔV.) Hence, with increasing pressure dP/dT approaches infinity and may become negative if ΔV becomes negative.

(b) <u>Standard states at P bar and T</u>

If we were to calculate the reaction curve shown in Fig. 2.1 using standard states of pure components at P bar and T, then the following procedure would be adopted. Since the standard state is in each case the pure phase at the P and T of interest and we are only considering equilibrium between pure phases, each component is always in its standard state. Therefore

$$(\mu_{CaSiO_3}^{woll})_{P,T} = \mu_{CaSiO_3}^{0} \qquad (2.33)$$

and the condition for equilibrium of the components involved in reaction (2.23) is

$$(\mu_{CaSiO_3}^{woll})_{P,T} + (\mu_{CO_2}^{fluid})_{P,T} - (\mu_{CaCO_3}^{cc})_{P,T} - (\mu_{SiO_2}^{qz})_{P,T}$$
$$= \Delta G = 0$$
$$(2.34)$$
$$\mu_{CaSiO_3}^{0} + \mu_{CO_2}^{0} - \mu_{CaCO_3}^{0} - \mu_{SiO_2}^{0} = \Delta G^0 = 0.$$

From (2.17) the following expression for μ_i^0 is obtained in this case :

$$\mu_i^0 = (H_j)_{1,T} - (T(S_j)_T + \int_1^P V_j \, dP. \tag{2.35}$$

Substituting eqn (2.35) for each of the components involved in the reaction into eqn (2.34) gives :

$$(H_{woll})_{1,T} + (H_{CO_2})_{1,T} - (H_{cc})_{1,T} - (H_{qz})_{1,T} -$$

$$- T \left((S_{woll})_T + (S_{CO_2})_T - (S_{cc})_T - (S_{qz})_T \right) + \tag{2.36}$$

$$+ \int_1^P (V_{woll} + V_{CO_2} - V_{qz} - V_{cc}) \, dP = 0,$$

or

$$\Delta H_{1,T} - T\Delta S_T + \int_1^P \Delta V \, dP = 0. \tag{2.37}$$

Making the assumptions that the volumes of the solids are independent of pressure and temperature and that CO_2 behaves as a perfect gas, eqn (2.37) becomes

$$\Delta H_{1,T} - T\Delta S_T + (P-1) \Delta V_{solids} + RT \ln P = 0. \tag{2.38}$$

Eqn (2.38) is exactly the same as eqns (2.29) and (2.31), demonstrating that the change of standard state has no effect on the result. All that has been done by changing the standard state from 1 bar, T to P bar, T is to remove the volume terms from the activity side of eqn (2.26) (Van't Hoff Isotherm) and include them in the standard state thermodynamic properties.

2.7 MORE ABOUT THE VAN'T HOFF ISOTHERM

Reaction (2.23) discussed above has the condition of equilibrium

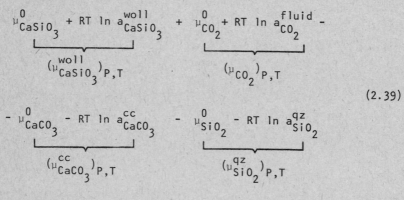

$$= 0.$$

Rearranging eqn (2.39) gives the Van't Hoff isotherm :

$$\Delta G^0 = - RT \ln \left(\frac{a_{CaSiO_3}^{woll} \cdot a_{CO_2}^{fluid}}{a_{CaCO_3}^{cc} \cdot a_{SiO_2}^{qz}} \right) \tag{2.40}$$

where ΔG^0 is the difference between chemical potentials of reactants and products in their standard states (e.g. 1 bar, T; P bar, T; 1 bar, 298 etc.) and a_i^j is the activity of component i in phase j under the pressure-temperature-composition conditions of interest. The activity term in brackets on the right-hand side of (2.40) is known as the equilibrium constant, K, of the reaction :

$$K = \left(\frac{a_{CaSiO_3}^{woll} \cdot a_{CO_2}^{fluid}}{a_{CaCO_3}^{cc} \cdot a_{SiO_2}^{qz}} \right). \tag{2.41}$$

Given values of standard state thermodynamic properties of the components involved in reaction (2.23), an equilibrium boundary in P-T space may be calculated from eqn (2.40) in the manner already discussed. An equilibrium curve of this type may be constructed for coexisting pure calcite, wollastonite, quartz, and CO_2, or as is usual in rocks, for the case of coexistence of impure multicomponent phases. The additional data needed in order to construct an equilibrium curve for impure phases are the compositions of the phases involved and the relationships between activities of components and phase compositions (see Chapter 3).

Relationships analogous to (2.40) and (2.41) may be written for any equilibrium which is relevant to the assemblage under consideration. In the general case of equilibrium between components A, B, C, and D the reaction of interest is

$$wA \ + \ xB \ \rightleftarrows \ yC \ + \ zD \tag{2.42}$$
$$\text{phase a} \qquad \text{b} \qquad \text{c} \qquad \text{d}$$

where w, x, y, z refer to the number of moles of components involved in the reaction.

At equilibrium we have :

$$w\mu_A^a + x\mu_B^b = y\mu_C^c + z\mu_D^d$$

$$w\mu_A^0 + w\ RT\ \ln\ a_A^a + x\mu_B^0 + x\ RT\ \ln\ a_B^b = y\mu_C^0 + y\ RT\ \ln\ a_C^c +$$

(2.43)

$$+ z\mu_D^0 + z\ RT\ \ln\ a_D^d.$$

Defining ΔG^0 as follows :

$$\Delta G^0 = y\mu_C^0 + z\mu_D^0 - w\mu_A^0 - x\mu_B^0, \tag{2.44}$$

eqn (2.43) may be rearranged to give

$$\Delta G^0 = -\ RT\ \ln\left(\frac{(a_C^c)^y \ . \ (a_D^d)^z}{(a_A^a)^w \ . \ (a_B^b)^x}\right). \tag{2.45}$$

In this case the equilibrium constant K is given by

$$K = \left(\frac{(a_C^c)^y \ . \ (a_D^d)^z}{(a_A^a)^w \ . \ (a_B^b)^x}\right). \tag{2.46}$$

Since in general the standard state of a component is taken to be the _pure_ phase at some pressure and temperature, μ^0 may usually be replaced by the molar free energy of the one-component phase under the standard state P-T conditions. For reaction (2.42), if the standard state of component A is taken to be pure A in phase a at some P and T, we have

$$\mu_A^0 = G_a^0 \quad \text{etc.} \tag{2.47}$$

2.8 IMPERFECT GASES

Taking the standard state of a gaseous species i to be the pure gas at P^0 bar and T, the activity of i in the pure gas at P bar and T is given by

$$RT \ln a_i = \int_{P^0}^{P} V_i \, dP. \tag{2.48}$$

If the gas is perfect, a_i is equal to the ratio (P/P^0) of pressure to standard state pressure (eqn (2.15)). However, because of intermolecular forces, real gases do not obey the equation of state (2.9) except at very low pressures. Eqn (2.15) does not, therefore, apply in the general case and it is necessary to know the actual pressure-volume-temperature (P-V-T) relationships of the real gas in order to calculate $(\mu_i)_{P,T}$ and a_i (see Appendix 1).

By definition the right-hand side of equation (2.48) for a pure gas gives the following result when integrated between P^0 and P :

$$\int_{P^0}^{P} V_i \, dP = RT \ln \left(\frac{f}{f^0}\right) \tag{2.49}$$

where f is the __fugacity__ (or thermodynamic pressure) of the gas at P bar and f^0 is its fugacity at P^0 bar. The fugacity of i in the pure gas is related to the pressure on the

gas P, as can be seen by comparing eqns (2.49) and
(2.15) :

$$f_i = P \ \Gamma_i; \ f_i^0 = P^0 \ \Gamma_i^0 \tag{2.50}$$

The variable Γ_i is the <u>fugacity coefficient</u> of the
gas i at the pressure and temperature of interest,
i.e. at P and T for f_i and at P^0 and T for f_i^0. At low
pressure intermolecular distances become large, intermolecular forces correspondingly small, and real gases
approach the perfect gas model in their properties. Under
these circumstances Γ_i approaches 1.0 and f_i approaches P.

2.9 CALCULATION OF REACTION INVOLVING IMPERFECT GAS

The 1 bar enthalpy and entropy changes of the muscovite
breakdown reaction,

$$\underset{\text{muscovite}}{KAl_3Si_3O_{10}(OH)_2} \rightleftarrows \underset{\text{sanidine}}{KAlSi_3O_8} + \underset{\text{corundum}}{Al_2O_3} + \underset{\text{fluid}}{H_2O}, \tag{2.51}$$

have been determined from a phase equilibrium study by
Chatterjee and Johannes (1974). In the temperature range
$600-800^{\circ}C$ these authors found that, within experimental
error, the 1 bar enthalpy and entropy changes of reaction
(2.51) can be assumed constant at the following values :

$$\Delta H_{1,T}^0 = + 23 \ 460 \ \text{cal} \ (98 \ 157 \ J);$$

$$\Delta S_T^0 = + 39.2 \ \text{e.u.} \ (164.0 \ J \ K^{-1} \ mol^{-1}).$$

The condition of equilibrium among the components in (2.51) is

$$\Delta G^0 = - RT \ln \left(\frac{a_{KAlSi_3O_8}^{san} \cdot a_{Al_2O_3}^{cor} \cdot a_{H_2O}^{fl}}{a_{KAl_3Si_3O_{10}(OH)_2}^{musc}} \right).$$

In petrological problems it is common to use the mixed standard state of P bar and T for solid components and 1 bar and T for fluid components. In this case ΔG^0 contains 1 bar, T data for all components and PV terms for solids only (eqns (2.27) and (2.37)) :

$$\Delta G^0 = \Delta H^0_{1,T} - T \Delta S^0_T + (P-1) V_{san} +$$
$$+ (P-1) V_{cor} - (P-1) V_{musc}.$$

(2.52)

If all phases are pure, then the solid components (with P, T standard state) are always in their standard state ($a_i = 1$) and (2.52) becomes

$$\Delta G^0 = \Delta H^0_{1,T} - T \Delta S^0_T + (P-1) \Delta V^0_{solids} = - RT \ln a_{H_2O}. \quad (2.53)$$

If H_2O behaves as a perfect gas, its activity relative to the 1 bar, T standard state is easy to determine (see eqn (2.15)). In practice, however, the P-V-T relationships for gaseous H_2O differ markedly from the perfect gas laws so that the simple equation of state (2.9) can only be applied at very low pressures. In this case the activity of H_2O at P bar and T is given by eqns (2.48) and (2.49) :

$$RT \ln a_{H_2O} = RT \ln \left(\frac{f_{H_2O}}{f^0_{H_2O}} \right) = RT \ln \left(\frac{P\Gamma_{H_2O}}{P^0\Gamma^0_{H_2O}} \right). \qquad (2.54)$$

Burnham et al. (1969) tabulate values of f_{H_2O} and Γ_{H_2O} for H_2O at temperatures of 20-1000°C and pressures of 100 to 10 000 bar (Appendix 1). Except at low temperatures (< 200°C), $\Gamma^0_{H_2O}$ at 1 bar is very close to 1.0 so that (with P^0 equal to 1.0) the activity of water becomes equal to its fugacity :

$$RT \ln a_{H_2O} = RT \ln f_{H_2O} = RT \ln P\Gamma_{H_2O}. \qquad (2.55)$$

The tabulated values of f_{H_2O} may therefore be used to determine a_{H_2O} at any desired pressure and temperature.

As an example we shall calculate the temperature at which pure muscovite, sanidine, corundum, and H_2O (fluid) would coexist at 3000 bar ($P_{H_2O} = P_{total}$). Substituting the values of $\Delta H^0_{1,T}$, ΔS^0_T, and ΔV^0_{solids} (= - 0.1515 cal bar^{-1}) given by Chatterjee and Johannes into eqn (2.53) gives

$$23\ 460 - T\ (39.2) - (2999)\ 0.1515 = - RT \ln a_{H_2O}. \qquad (2.56)$$

Rearranging (2.56) and assuming $a_{H_2O} = f_{H_2O}$ yields

$$- \frac{23\ 460}{RT} + \frac{39.2}{R} + \frac{2999 \times 0.1515}{RT} = \ln f_{H_2O}. \qquad (2.57)$$

Eqn (2.57) cannot be solved directly for T since f_{H_2O} is not equal to P_{H_2O} at 3000 bar pressure. It may, however, be solved in a few minutes by successive approximations of the equilibrium temperature. If we substitute T = 973 K (700°C) into eqn (2.57), then the calculated equilibrium f_{H_2O} is 2515 bar. In the tables of Burnham et al., f_{H_2O} of 2515 bar corresponds to P_{H_2O} of 3550 bar. This means that at 973 K, a P_{H_2O} of greater than 3000 bar would be required to stabilize muscovite in the presence of sanidine, corundum, and water. Therefore, at P_{H_2O} = 3000 bar, muscovite is unstable at 700°C. Since muscovite has lower entropy than the products of reaction (2.51), the equilibrium temperature must be less than 700°C. At 923 K (650°C), the equilibrium f_{H_2O} calculated from (2.57) is 1320 bar, and P_{H_2O} = 2250 bar. At 3000 bar, 650°C, therefore, muscovite is more stable than sanidine, corundum, and water. The equilibrium temperature at 3000 bar assuming $P_{H_2O} = P_{total}$ lies, therefore, between 650 and 700°C.

Two further temperature approximations are sufficient to locate the equilibrium boundary within experimental error :

943 K (670°C) f_{H_2O} = 1722 bar; P_{H_2O} = 2800 bar

953 K (680°C) f_{H_2O} = 1960 bar; P_{H_2O} = 3080 bar.

Linear interpolation between these two results fixes the equilibrium temperature as 950 K, (677°C) at a P_{H_2O} of 3000 bar.

For a more detailed discussion of standard states in calculation of dehydration equilibria, the reader is referred to Anderson (1970).

SUMMARY - CHAPTER 2

1.   The chemical potential of component i in phase j under a given set of conditions μ_i^j (pressure, temperature, composition) is referred to a <u>standard state</u> value μ_i^0 by the equation

$$\mu_i^j = \mu_i^0 + RT \ln a_i^j \qquad (2.58)$$

where a_i^j is the activity of component i in phase j under the conditions of interest.

2.   If phase j is pure, consisting only of component i, then the activity of i in eqn (2.58) depends on the standard state chosen :

(a) standard state of pure i (phase j) at the pressure P and temperature T of interest :

$$RT \ln a_i^j = 0$$

$$a_i^j = 1;$$

(b) standard state of pure i in j at pressure P^0 and the temperature of interest :

$$RT \ln a_i^j = \int_{P^0}^{P} V_j \, dP. \qquad (2.59)$$

3.   The standard states used for any particular calculation are generally those which are most convenient and which facilitate manipulation of the available thermodynamic data. The actual value of μ_i^j under the conditions of interest is independent of the standard state chosen.

4.   For a perfect gas, eqn (2.59) may be integrated to give

$$RT \ln a_i^j = RT \ln (P/P^0).$$

If the gas is imperfect, then the integration of (2.59) yields the ratio of fugacities (f) of the gas at the pressure of interest and at the standard state pressure :

$$RT \ln a_i^j = RT \ln \left(\frac{f_i}{f_i^0} \right). \qquad (2.60)$$

Fugacity is related to pressure by

$$f_i = P\Gamma_i.$$

Γ_i is a coefficient (fugacity coefficient) which is dependent on pressure and temperature. Values of Γ_i are known (approximately) for many of the pure gases of geological interest.

5.   For a reaction of the type

$$wA + xB \rightleftarrows yC + zD$$

phase   a   b   c   d

(2.61)

the condition of equilibrium is

$$(\Delta G^0) = - RT \ln \left(\frac{(a_C^c)^y \cdot (a_D^d)^z}{(a_A^a)^w \cdot (a_B^b)^x} \right)$$

(2.62)

where (ΔG^0) is the difference between the sums of standard-state chemical potentials of product and reactant components and a_D^d refers to the activity of component D in phase d under the pressure-temperature-composition conditions of interest.

PROBLEMS - CHAPTER 2

1.   Calculate the activity of the $Mg_3Al_2Si_3O_{12}$ component in pure pyrope ($Mg_3Al_2Si_3O_{12}$) garnet at 10 kbar and 800°C, taking the standard state of the component to be pure pyrope at 1 bar and 800°C.

V_{pyrope} at 1 bar, 298 K = 113.29 cm^3 = 2.7077 cal bar^{-1}
(assume constant).

2.   The reaction

$Mg(OH)_2 \rightleftharpoons MgO + H_2O$
brucite   periclase   fluid

has the following approximate values of $\Delta H^0_{1,T}$, ΔS^0_T at
1 bar, 900 K, and ΔV^0_{solids} at 1 bar, 298 K :

$\Delta H^0_{1,T}$ = 15 100 cal; S^0_T = 29.6 e.u.

ΔV^0_{solids} = - 0.3198 cal bar^{-1} (assume constant).

Assuming ΔC_p = 0, calculate the equilibrium temperature
for pure brucite, pure periclase and pure H_2O-fluid at a
pressure of 1000 bar.   Perform the calculation for the
two cases in which :

(a) H_2O behaves as a perfect gas;

(b) Γ_{H_2O} at 1000 bar = 0.65; $\Gamma^0_{H_2O}$ at 1 bar is 1.0.

SOLUTIONS TO PROBLEMS - CHAPTER 2

1.   The activity of a component in a pure one-component
phase at any pressure and temperature, taking the
standard state at 1 bar and the temperature of interest,
is given by eqn (2.7) :

$$RT \ln a^{pyrope}_{Mg_3Al_2Si_3O_{12}} = \int_1^P V_{pyrope} \, dP.$$

Assuming V independent of P and T this becomes

$$RT \ln a^{pyrope}_{Mg_3Al_2Si_3O_{12}} = (P-1) \times 2.7077.$$

Substituting P = 10 000 and T = 1073 ($800^\circ C$),

$$\ln a^{pyrope}_{Mg_3Al_2Si_3O_{12}} = \frac{9999 \times 2.7077}{1.987 \times 1073}$$

$$a^{pyrope}_{Mg_3Al_2Si_3O_{12}} = \underline{3.273 \times 10^5}.$$

2.   The equilibrium condition for products and reactants
of the reaction

$$\begin{array}{ccc} Mg(OH)_2 & \rightleftharpoons & MgO & + & H_2O \\ \text{brucite} & & \text{periclase} & & \text{fluid} \end{array}$$

is

$$\Delta G^0 = - RT \ln \left(\frac{a^{periclase}_{MgO} \cdot a^{fluid}_{H_2O}}{a^{brucite}_{Mg(OH)_2}} \right).$$

Taking standard states of pure solids at the pressure
and temperature of interest ($a^{periclase}_{MgO}$ = 1 and
$a^{brucite}_{Mg(OH)_2}$ = 1), this equation becomes

$$\Delta H^0_{1,T} - T\Delta S^0_T + (P-1)\ \Delta V^0_{solids} = -\ RT\ \ln a_{H_2O}.$$

(a) Assuming that H_2O is a perfect gas with 1 bar, T standard state, a_{H_2O} = P/1 and we obtain

$$15\ 100 - T\ (29.6) + 999\ (-\ 0.3198) = -\ RT\ \ln 1000$$
$$14\ 781 = 15.874\ T$$
$$T = \underline{931\ K}.$$

(b) For non-ideal gases

$$a_i = \frac{P\Gamma}{P^0\Gamma^0}.$$

$\Gamma^0 = 1.0, \Gamma = 0.65$ and $P/P^0 = 1000$ in this case.

Hence we obtain

$$15\ 100 - T\ (29.6) + 999\ (-\ 0.3198) = -\ RT\ \ln$$
$$(0.65 \times 1000)$$
$$14\ 781 = 16.73\ T$$
$$T = \underline{883\ K}.$$

It is apparent that in this case the assumption that H_2O behaves as a perfect gas at high pressures is not adequate to enable accurate calculation of the equilibrium boundary.

3. Multicomponent solids and fluids

3.1 INTRODUCTION

The calculation of equilibrium conditions using thermodynamic principles has thus far only been applied to
assemblages of solid and fluid phases which are pure. In
rocks, of course, mineral, fluid and liquid phases are
never pure and seldom even approximate to one-component
phases. It is apparent, therefore, that thermodynamic
calculations for pure phases are only applicable to rocks
in a qualitative way and that it is essential to understand
the influence of mixing in multicomponent phases before
quantitative calculations can be made. Fortunately there
is an increasing body of data on the thermodynamic
properties of multicomponent solid and liquid silicate
phases and on mixed-fluid phases. In cases where little
information is available on mixing properties, it is often
possible to make reasonable approximations which enable
quantitative calculations to be made. This chapter is
devoted to a discussion of the types of model which can be
used to estimate the thermodynamic properties of multicomponent solid and fluid phases and to some of the available

results on multicomponent phases. A large part of the
discussion is applicable to multicomponent liquids as well
as solids and fluids, but the former will be dealt with in
more detail in Chapter 5.

3.2 ENTROPY OF MIXING

Consider one mole of a solid solution between MgO
(periclase) and FeO (wüstite). The entropy of the (Mg,Fe)O
solid solution may be considered to be made up of the
thermal and mixing contributions discussed briefly in
Chapter 1. Provided that the structures and volumes of
the two pure end-members (in this case MgO and FeO) and
the solid solution are similar, the thermal contribution
to the molar entropy should differ little from the sum of
the thermal entropies of the two end-members, i.e.

$$S_{th} = X_{MgO} \, S_{per} + X_{FeO} \, S_{wu} \qquad (3.1)$$

where S_{th} is the thermal contribution to the molar entropy,
S_{wu} and S_{per} are the molar entropies of pure FeO and MgO
respectively, and the X's are mole fractions in the solid
solution. (Note that in pure FeO and pure MgO the entropy
is solely thermal.)

The entropy-of-mixing contribution to the entropy of
the (Mg,Fe)O solid solution is given by the Boltzmann
relationship

$$S_{mix} = k \ln Q \qquad (3.2)$$

where S_{mix} is the entropy of mixing, k is the Boltzmann
constant (gas constant, R, divided by Avogadro's number
A), and Q (the permutability) is the total number of
possible atomic configurations in the mixture.

The total number of possible atomic configurations
in the (Mg,Fe)O mixture may be readily calculated from the
mathematical formula for combinations. Suppose we have N
positions filled by N_{Fe} atoms of iron (indistinguishable
from one another) and N_{Mg} atoms of magnesium (distinguishable
from iron but not from each other).   If the N_{Fe} and N_{Mg}
atoms are randomly distributed over the N positions, then
the total number of possible configurations is given by :

$$Q = \frac{N!}{N_{Fe}! \, N_{Mg}!} \tag{3.3}$$

where ! designates factorial, (N. (N-1). (N-2) ... 1).
Substituting eqn (3.3) into (3.2) gives

$$S_{mix} = k \ln \frac{N!}{N_{Fe}! \, N_{Mg}!} = k \ln N! - k \ln N_{Fe}!$$
$$- k \ln N_{Mg}! \tag{3.4}$$

Since the numbers of sites and atoms involved in
mixing are very large (multiples of Avogadro's number,
6.02×10^{23} per mole),Stirling's approximation,

$$\ln N! = N \ln N - N \text{ (for large numbers)},$$

may be applied to the factorials in eqn (3.4). This yields

$$S_{mix} = k \; (N \ln N - N_{Fe} \ln N_{Fe} - N_{Mg} \ln N_{Mg}). \qquad (3.5)$$

Applying the constraint that N is equal to the sum of N_{Mg} and N_{Fe} and collecting terms together results in

$$S_{mix} = - \; k \; N \left[\left(\frac{N_{Fe}}{N_{Fe} + N_{Mg}} \right) \ln \left(\frac{N_{Fe}}{N_{Fe} + N_{Mg}} \right) \right.$$
$$\left. + \left(\frac{N_{Mg}}{N_{Fe} + N_{Mg}} \right) \ln \left(\frac{N_{Mg}}{N_{Fe} + N_{Mg}} \right) \right] \qquad (3.6)$$

The fraction of the cation positions in the (Mg,Fe)O solid solution occupied by iron atoms is

$$X_{Fe} = \left(\frac{N_{Fe}}{N_{Fe} + N_{Mg}} \right). \qquad (3.7)$$

Therefore, if we consider one mole (Avogadro's number of formula units) of the solid solution, eqn (3.6) becomes

$$S_{mix} = - \; n \; R \; (X_{Fe} \ln X_{Fe} + X_{Mg} \ln X_{Mg}). \qquad (3.8)$$

In eqn (3.8), n is the number of positions in each formula unit on which mixing takes place and R is the gas constant. In the case of (Mg,Fe)O solution, n is equal to 1. In other solid solutions n may generally be obtained from a consideration of the formula of the phase, e.g.

$(Fe,Mg)_2SiO_4$ olivine $n = 2$ (assuming no mixing on Si positions);

$(Ca,Mg,Mn,Fe)_3Al_2Si_3O_{12}$ garnet $n = 3$ (assuming no mixing on Al or Si positions);

$Ca(Mg,Fe,Mn)Si_2O_6$ pyroxene $n = 1$ (no mixing on Ca or Si positions).

If the sites in the phase on which mixing takes place are occupied by more than two types of atom, then the formula for the entropy of mixing (eqn (3.8)) may be extended using fractions of all atoms present. For example, if we were to consider a $(Ca,Mg,Fe,Mn)_3Al_2Si_3O_{12}$ garnet, the entropy of mixing would be

$$S_{mix} = - 3R (X_{Ca} \ln X_{Ca} + X_{Mg} \ln X_{Mg} + X_{Fe} \ln X_{Fe} + X_{Mn} \ln X_{Mn}) \tag{3.9}$$

where

$$X_{Ca} = \left(\frac{Ca}{Ca + Mg + Fe + Mn} \right),$$

$$X_{Mg} = \left(\frac{Mg}{Ca + Mg + Mn + Fe} \right) \text{ etc.}$$

Eqn (3.9) may be generalized to the case of i distinct species mixing on n positions per formula unit as follows :

$$S_{mix} = - n R \left(\sum_i X_i \ln X_i \right). \tag{3.10}$$

3.3   FREE ENERGY OF AN IDEAL SOLUTION

Consider a system made up of X_A moles of the oxide A_nO and X_B moles of the oxide B_nO (where $X_A + X_B = 1$) held at some pressure P and temperature T. If each phase is pure, then, taking the standard state of each component to be the pure phase at P and T, the total free energy of the system is given by

$$G^0 = X_A \, \mu^0_{A_nO} + X_B \, \mu^0_{B_nO} \qquad\qquad (3.11)$$

Suppose now that A_nO and B_nO are allowed to mix to form one homogeneous solid solution $(A,B)_nO$ with randomly mixed atoms of A and B on the n cation positions per formula unit. If the $(A,B)_nO$ solid solution is ideal, the only change in the free energy of the system is due to the entropy of A-B mixing and there is no heat of mixing ($H_{mix} = 0$). In this case the free energy of the solid solution (G_{ss}) is given by

$$G_{ss} = \underbrace{X_A \, \mu^0_{A_nO} + X_B \, \mu^0_{B_nO}}_{G^0} + \underbrace{n \, RT \, (X_A \ln X_A + X_B \ln X_B)}_{G_{mix}}. \quad (3.12)$$

The G_{mix} contribution to G_{ss} is given by

$$G_{mix} = H_{mix} - TS_{mix} = -\,TS_{mix}. \qquad\qquad (3.13)$$

3.4   ACTIVITIES OF COMPONENTS IN IDEAL SOLUTIONS

In order to calculate equilibria using the methods discussed in Chapters 1 and 2, it is necessary to determine the chemical potentials or activities of components in the multicomponent phases occurring in nature.

An ideal solid solution of the type $(A,B)_n O$ has a molar free energy given by eqn (3.12). The free energy of a system or phase can also be expressed by the sum of the chemical potentials of its constituent components at the composition, times the number of moles of each component (eqn (1.4)).

In this case :

$$G_{ss} = X_A \, \mu_{A_n O} + X_B \, \mu_{B_n O}. \qquad\qquad (3.14)$$

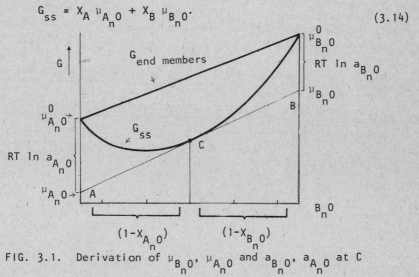

FIG. 3.1.   Derivation of $\mu_{B_n O}$, $\mu_{A_n O}$ and $a_{B_n O}$, $a_{A_n O}$ at C from G_{ss} and $\partial G_{ss}/\partial X_A$ (represented by the line ACB) - see text.

Substituting $(1-X_A)$ for X_B in eqn (3.14) and differentiating with respect to, X_A, it may be shown that $\mu_{A_n O}$ in the solid solution is given by

$$\mu_{A_n O} = G_{ss} + (1-X_A) \left(\frac{\partial G_{ss}}{\partial X_A} \right)_{P,T} . \qquad (3.15)$$

Eqn (3.15) is the general equation for the derivation of partial molar quantities from the molar properties of any binary solution. In this case it is being used to obtain the partial molar free energy (chemical potential) of one component from the total free energy of the phase.

A graphical illustration of the derivation of $\mu_{A_n O}$ and $\mu_{B_n O}$ from values of G_{ss} and $(\partial G_{ss}/\partial X_A)$ is given in Fig. 3.1. From (3.15) the tangent at C, when extrapolated to the points $X_A = 1$ and $X_A = 0$ gives, respectively, $\mu_{A_n O}$ and $\mu_{B_n O}$ at composition C.

Expressions analogous to (3.15) for partial molar volumes, enthalpies etc. may be written as follows :

$$\bar{V}_{A_n O} = V_{ss} + (1-X_A) \left(\frac{\partial V_{ss}}{\partial X_A} \right)_{P,T}$$

$$\bar{H}_{A_n O} = H_{ss} + (1-X_A) \left(\frac{\partial H_{ss}}{\partial X_A} \right)_{P,T} . \qquad (3.16)$$

Example

X-ray studies of albite ($NaAlSi_3O_8$) - sanidine ($KAlSi_3O_8$) solid solutions have provided unit cell

dimensions from which Waldbaum and Thompson (1968) have
obtained the following equation for molar volumes :

$$V_{ss} = 2.394 + 0.2978 \ X_{or} - 0.0894 \ X_{or}^2 \ \text{cal bar}^{-1}$$

where X_{or} is the mole fraction of orthoclase component.
Differentiating with respect to X_{or},

$$\left(\frac{\partial V_{ss}}{\partial X_{or}}\right) = 0.2978 - 0.1788 \ X_{or}.$$

Comparison with eqn (3.16) and Fig. 3.1 gives the following
expressions for $\bar{V}_{or}$ and $\bar{V}_{alb}$:

$$\bar{V}_{or} = \underbrace{2.394 + 0.2978 \ X_{or} - 0.0894 \ X_{or}^2}_{V_{ss}} + (1-X_{or}) \ (0.2978 - 0.1788 \ X_{or})$$

$$\bar{V}_{alb} = \underbrace{2.394 + 0.2978 \ X_{or} - 0.0894 \ X_{or}^2}_{V_{ss}} - (1-X_{or}) \ (0.2978 - 0.1788 \ X_{or})$$

The activity of component $A_n O$ in the $(A,B)_n O$ solid solution
may now be obtained by substituting $(1-X_A)$ for X_B in eqn
(3.12) and differentiating with respect to X_A :

$$\left(\frac{\partial G_{ss}}{\partial X_A}\right)_{P,T} = \mu_{A_n O}^0 - \mu_{B_n O}^0 + n \ RT \ (\ln X_A - \ln (1-X_A)). \qquad (3.17)$$

Multiplying the right-hand side of (3.17) by $(1-X_A)$ and substituting into (3.15) gives

$$\mu_{A_nO} = G_{ss} + (1-X_A)\ \mu^0_{A_nO} - (1-X_A)\ \mu^0_{B_nO} +$$

$$+ n\ RT\ (1-X_A)\ (\ln X_A - \ln (1-X_A)).$$

Substituting for G_{ss} (eqn (3.12)), we obtain

$$\mu_{A_nO} = \mu^0_{A_nO} + n\ RT\ \ln X_A. \tag{3.18}$$

Consider now the expression for the activity of component i :

$$\mu_i = \mu^0_i + RT\ \ln a_i. \tag{3.19}$$

From this it can be seen that the activity of the component A_nO in the solid solution, <u>taking the standard state of A_nO as pure A_nO at the pressure and temperature of interest</u>, is given by

$$a_{A_nO} = X_A^n. \tag{3.20}$$

Eqn (3.20) applies for any ideal solid solution containing n sites per formula unit on which mixing takes place and where there is only one distinct type of site involved (in this case the A, B sites). Although derived above for a simple binary solid solution, the result is applicable to solid solutions with any number of components, e.g.

(a) $(Ca,Fe,Mn,Mg)_3Al_2Si_3O_{12}$ garnet.   If ideal,

$$a_{Ca_3Al_2Si_3O_{12}} = X_{Ca}^3$$

where

$$X_{Ca} = \left(\frac{Ca}{Ca + Fe + Mn + Mg} \right)$$

in the eight coordinate cubic garnet site.

(b) $Ca(Mg,Fe,Mn)Si_2O_6$ clinopyroxene.   If ideal,

$$a_{CaFeSi_2O_6} = X_{Fe}^1$$

where

$$X_{Fe} = \left(\frac{Fe}{Fe + Mg + Mn} \right)$$

in the clinopyroxene M1 site.

3.5 CALCULATIONS OF EQUILIBRIA INVOLVING IDEAL SOLUTIONS

The coexistence of plagioclase, garnet, sillimanite, and quartz is controlled by the equilibrium

$$3CaAl_2Si_2O_8 \rightleftharpoons Ca_3Al_2Si_3O_{12} + 2Al_2SiO_5 + SiO_2. \quad (3.21)$$
plagioclase   garnet   sillimanite   quartz

Taking standard states of all components to be the pure

phase at the pressure and temperature of interest, the
standard free energy change of this reaction is given by

$$(\Delta G^0)_{P,T} = \Delta H^0_{1,T} - T\Delta S^0_T + (P-1) \Delta V^0_{solids}$$

$$= -10\ 300 + 31.83T - 1.3045\ (P-1)\ cal$$
$$(= -43\ 095 + 133.18T - 5.458\ (P-1)\ J).$$

Equilibrium in any assemblage of these four phases,
regardless of complexity, necessitates the condition

$$(\Delta G^0)_{P,T} = -RT\ \ln\left(\frac{a^{gt}_{Ca_3Al_2Si_3O_{12}} \cdot a^{sill^2}_{Al_2SiO_5} \cdot a^{qtz}_{SiO_2}}{a^{plag^3}_{CaAl_2Si_2O_8}}\right). \quad (3.22)$$

Let us suppose that all phases in such an assemblage
are pure, consisting solely of the components given in
reaction (3.21). The activity of every component using
the P, T standard state, is equal to 1.0, so that (3.22)
becomes

$$(\Delta G^0)_{P,T} = -RT\ \ln 1 = 0$$
$$= -10\ 300 + 31.83T - 1.3045\ (P-1). \quad (3.23)$$

This equation may now be solved to calculate a P-T
line along which the four pure phases would be in equilibrium. At, for example, 900 K,

$$P = \frac{-10\ 300 + 900 \times 31.83}{1.3045} + 1 = \underline{14\ 065\ bar}.$$

At 1200 K,

$$P = \frac{-10\ 300 + 1200 \times 31.83}{1.3045} + 1 = \underline{21\ 385\ bar}.$$

In rocks which contain this assemblage, however, not all of the phases are pure, so that these results do not necessarily apply. Generally, sillimanite is close to pure Al_2SiO_5 and quartz is almost pure SiO_2, but natural plagioclase is a $CaAl_2Si_2O_8$ - $NaAlSi_3O_8$ solid solution and natural garnet is a solid solution of approximate composition $(Ca,Fe^{2+},Mg,Mn)_3Al_2Si_3O_{12}$.

Consider a rock containing pure quartz, pure sillimanite, plagioclase of 0.2 $X_{CaAl_2Si_2O_8}$, and garnet with $X_{Ca} = (Ca/(Ca + Fe^{2+} + Mg + Mn))$ equal to 0.05. If the garnet solid solution is ideal, the activity of the $Ca_3Al_2Si_3O_{12}$ component is given by

$$a^{gt}_{Ca_3Al_2Si_3O_{12}} = (Ca/(Ca + Fe^{2+} + Mg + Mn))^3 = 0.05^3. \quad (3.24)$$

Although mixing in $NaAlSi_3O_8$ - $CaAl_2Si_2O_8$ plagioclase takes place on more than one site per formula unit, it has been found experimentally, probably because of charge-balance constraints (see section 3.6), that activity is approximately equal to mole fraction :

$$a^{plag}_{CaAl_2Si_2O_8} = X^{plag}_{CaAl_2Si_2O_8} = 0.2.$$

Substituting these two activity expressions into eqn (3.22) gives

$$- 10\ 300 + 31.83T - 1.3045\ (P-1) = - RT \ln \left(\frac{0.05^3}{0.2^3} \right)$$
$$- 3\ RT \ln \left(\frac{0.05}{0.2} \right).$$

This equation may now be solved to calculate a P-T line along which the complex assemblage would be in equilibrium. At 900 K,

$$P = \frac{- 10\ 300 + 900 \times 31.83 + 1.987 \times 3 \times 900 \times \ln 0.25}{1.3045}$$
$$+ 1 = \underline{8\ 364\ \text{bar}}.$$

At 1200 K,

$$P = \frac{- 10\ 300 + 1200 \times 31.83 + 1.987 \times 3 \times 1200 \times \ln 0.25}{1.3045}$$
$$+ 1 = \underline{13\ 784\ \text{bar}}.$$

It is apparent that the effect of solid solution in this case is to lower drastically the equilibrium pressure of the natural assemblage relative to the pure, simple system. The calculated P-T lines for pure and complex cases are shown in Fig. 3.2.

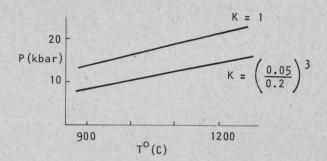

FIG. 3.2. The calculated equilibrium condition of pure
$CaAl_2Si_2O_8$ plagioclase, pure $Ca_3Al_2Si_3O_{12}$ garnet, pure
sillimanite, and pure quartz (line labelled K = 1), and
under which multicomponent phases would coexist (K =
$(0.05/0.2)^3$ - see text).

3.6 THE EFFECT OF LOCAL CHARGE BALANCE ON ACTIVITY

Consider a clinopyroxene solid solution $(Ca^{2+}, Na^+)^{M2}$
$(Al^{3+}, Mg^{2+})^{M1}$ Si_2O_6 where the superscripts M2 and M1
refer to the two distinct types of site in the clinopyroxene C2/c structure. If there is complete disorder
on both cation sites in the formula unit, then the 'ideal'
activities of $CaMgSi_2O_6$ and $NaAlSi_2O_6$ components are given
(by analogy with n equal to 2, see section 3.7) by

$$a^{cpx}_{CaMgSi_2O_6} = X^{M2}_{Ca} \cdot X^{M1}_{Mg}$$

$$a^{cpx}_{NaAlSi_2O_6} = X^{M2}_{Na} \cdot X^{M1}_{Al}$$

where

$$X_{Ca}^{M2} = \left(\frac{Ca}{Ca + Na} \right)_{M2}$$

$$X_{Al}^{M1} = \left(\frac{Al}{Al + Mg} \right)_{M1} \quad \text{etc.}$$

Since, however, Ca^{2+} and Na^+ ions and Al^{3+} and Mg^{2+} ions have different charges, it is apparent that complete disorder over both sites will produce regions of local charge imbalance in the structure. Thus, although $NaAlSi_2O_6$ and $CaMgSi_2O_6$ are both neutral 'molecules', complete disorder over both positions will produce regions having formulae '$NaMgSi_2O_6^-$' and '$CaAlSi_2O_6^+$' which are probably energetically unfavourable.   It seems likely, therefore, that there is a certain amount of coupling of Na and Al and Ca and Mg in the clinopyroxene structure to preserve charge balance locally. The effect of coupling is, of course, to decrease the configurational (mixing) entropy of the solid solution and to alter the activity-composition relation-ships.

Consider a clinopyroxene in which Al-Mg mixing on M1 is completely random but in which, whenever there is an Al atom on M1, a sodium atom resides in one of the three nearest-neighbour M2 positions to preserve local charge balance. If, as a limiting case, the charge balancing sodium atom always resides in the same nearest-neighbour M2, the M2 positions do not contribute to the entropy of the solid solution. This is because the positions of the sodium atoms are

completely fixed by the positions of the aluminium atoms.
Thus, the total entropy of mixing is simply that which arises
from disorder on M1 :

$$S_{mix} = - R (X_{Al}^{M1} \ln X_{Mg}^{M1} + X_{Al}^{M1} \ln X_{Mg}^{M1}), \tag{3.25}$$

and the activity-composition relations for the solid solution
(if ideal) are analogous to the one-site mixing case :

$$a_{NaAlSi_2O_6}^{cpx} = X_{Al}^{M1}$$

$$a_{CaMgSi_2O_6}^{cpx} = X_{Mg}^{M1}.$$

If the substituting sodium atom 'distinguishes' between the
different M2 sites in this way, then the overall symmetry
of the pyroxene should be lowered, since in the relatively
high symmetry space group C2/c all three M2 sites are
equivalent. This deduction is in accordance with the
observation that natural jadeitic clinopyroxenes (omphacites)
have lower symmetry (P2) than diopside (C2/c), and that the
M2 positions are split into four distinct types (Clark et al.
1969).

Ganguly (1973) has shown that activity-composition
relationships in natural and synthetic omphacitic clinopyroxenes closely approximate to the ordered one-site mixing
model described above. Similar relationships have been
observed in plagioclase (Bowen 1928, p. 176) and in aluminous
orthopyroxenes (Wood and Banno 1973). In plagioclase the
substitution is $Na^+Al^{3+}Si^{4+}$ for $Ca^{2+}Al^{3+}Al^{3+}$ and in ortho-

pyroxene $Al^{3+}Al^{3+}$ for $Mg^{2+}Si^{4+}$, but despite the multi-site mixing, both approximate to the locally-ordered ideal one-site mixing model :

$$a_i \simeq X_i.$$

3.7 MORE ABOUT MULTI-SITE MIXING

Although ideal activity-composition relationships for mixing on two different types of site in the formula unit have already been introduced with reference to (CaNa) $(AlMg)Si_2O_6$ clinopyroxene, it is necessary in view of the abundance of multi-site solid solutions in rocks to discuss further the derivation of ideal activities for these phases.

Consider a garnet solid solution between $Ca_3Fe_2^{3+}Si_3O_{12}$, $Ca_3Al_2Si_3O_{12}$, and $Mg_3Al_2Si_3O_{12}$ in which there is mixing on both the cubic and octahedral sites in the structure. Assuming ideal random mixing on both types of site, the molar free energy of the $(Ca,Mg)_3 (Al,Fe^{3+})_2 Si_3O_{12}$ solid solution is given by (standard states at P and T)

$$
\begin{aligned}
G_{ss} = {} & X_{Ca_3Fe_2Si_3O_{12}} \; \mu^0_{Ca_3Fe_2Si_3O_{12}} + \\
& + X_{Ca_3Al_2Si_3O_{12}} \; \mu^0_{Ca_3Al_2Si_3O_{12}} + \\
& + X_{Mg_3Al_2Si_3O_{12}} \; \mu^0_{Mg_3Al_2Si_3O_{12}} + 3\,RT\,(X^c_{Ca}\,\ln X^c_{Ca} + \\
& + X^c_{Mg}\,\ln X^c_{Mg}) + 2\,RT\,(X^{oct}_{Al}\,\ln X^{oct}_{Al} + X^{oct}_{Fe}\,\ln X^{oct}_{Fe}),
\end{aligned}
\tag{3.26}
$$

where superscripts c and oct refer respectively to cubic

and octahedral sites in the solid solution and X_{Al}^{oct} =
$(Al/(Al + Fe))_{oct}$; $X_{Ca}^c = (Ca/(Ca + Mg))_c$ etc.

It may readily be seen by comparison with eqn (3.18)
that the chemical potential of the $Mg_3Al_2Si_3O_{12}$ component
in this solid solution is given by

$$\mu_{Mg_3Al_2Si_3O_{12}}^{gt} = \mu_{Mg_3Al_2Si_3O_{12}}^0 + 3\ RT\ \ln X_{Mg}^c + 2\ RT\ \ln X_{Al}^{oct}$$

$$= \mu_{Mg_3Al_2Si_3O_{12}}^0 + RT\ \ln X_{Mg}^{c3} \cdot X_{Al}^{oct2} \qquad (3.27)$$

Comparison of (3.27) with (3.19) gives the activity of the
$Mg_3Al_2Si_3O_{12}$ component in the garnet solution :

$$a_{Mg_3Al_2Si_3O_{12}}^{gt} = X_{Mg}^{c3} \cdot X_{Al}^{oct2} . \qquad (3.28)$$

Results analogous to (3.28) may be applied to any solid
solution in which there is mixing on more than one type of
site and in which there is complete disorder of atoms on
each site. For example,

$$(Ca,Mg,Fe^{2+},Mn)_3(Al,Fe^{3+})_2Si_3O_{12};$$
$$\text{garnet}$$
$$a_{Ca_3Al_2Si_3O_{12}} = X_{Ca}^{c3} \cdot X_{Al}^{oct2}$$

where $X_{Ca} = (Ca/(Ca + Fe^{2+} + Mg + Mn)$ etc. (note that
there are no charge balance constraints in this case).

Similarly for a clinopyroxene :

$$(Ca,Na,Mg,Fe^{2+})_{M2} \ (Al,Fe^{2+},Mg)_{M1} \ (Si,Al)_2O_6.$$

Assuming complete disorder on M2, M1 and tetrahedral positions (no local charge balance), and ideal mixing on each type of site :

$$a^{cpx}_{CaMgSi_2O_6} = X^{M2}_{Ca} \cdot X^{M1}_{Mg} \cdot X^{tet^2}_{Si}$$

$$a^{cpx}_{NaAlSi_2O_6} = X^{M2}_{Na} \cdot X^{M1}_{Al} \cdot X^{tet^2}_{Si}$$

where

$$X^{M2}_{Na} = \left(\frac{Na}{Ca + Na + Mg_{M2} + Fe^{2+}_{M2}} \right)$$

and so on.

$(Mg,Fe)^{2+}_{M2} \ (Mg,Fe)^{2+}_{M1} \ Si_2O_6$ orthopyroxene (no charge balance constraints) has

$$a^{opx}_{Mg_2Si_2O_6} = X^{M2}_{Mg} \cdot X^{M1}_{Mg} \qquad (X^{M2}_{Mg} \neq X^{M1}_{Mg} \text{ in orthopyroxene}).$$

3.8 NON-IDEAL SOLUTIONS

In many cases the use of ideal solution relationships for silicate minerals enables reasonably accurate calculations of equilibria in complex systems to be made (see Chapter 4). Strictly, however, all silicate solid solutions

deviate to some extent from ideality because H_{mix} is never
exactly zero, and in many cases these deviations are so
large that miscibility gaps are formed (e.g. alkali-
feldspars, hemoᴛilmenites). It is often necessary, therefore, to take account of non-ideality when using activity-
composition relationships in thermodynamic calculations.
The next section is devoted to the simplest type of non-ideal
solution model (symmetrical regular solution) which has been
applied to a number of silicate solid solutions. In such a
solution the heat of mixing is not zero, but the entropy of
mixing is taken to be ideal.

It will be emphasized throughout this discussion that,
as with ideal solution models, non-ideal models involve a
number of assumptions which may or may not be valid in any
particular case. It is essential before using activity-
composition relationships derived using a simple model that
the reader appreciate the restrictions placed on his results
by these assumptions.

3.9 SYMMETRICAL REGULAR SOLUTIONS

The molar free energy of mixing of a binary regular
solution $(A,B)_n O$ is given by

$$G_{mix} = \underbrace{n\, RT\, (X_A \ln X_A + X_B \ln X_B)}_{G^{ideal}_{mix}} + \underbrace{n\, X_A X_B W_G}_{G^{xs}} \cdot \qquad (3.29)$$

The non-ideal contribution to the free energy, G^{xs}, is a
symmetrical function of composition :

$$G^{xs} = n \, X_A \, X_B \, W_G = n \, X_A \, (1-X_A) \, W_G \qquad (3.30)$$

and W_G is the A-B <u>interchange energy</u> or <u>interaction parameter</u>.
Note that it is assumed that the entropy of mixing is not
altered by the non-ideal interactions and that there is no
ordering or clustering of atoms.

Comparing eqn (3.30) with (3.12), it may be seen that the
total free energy of the regular solid solution is given by

$$G_{ss} = X_A \, \mu^0_{A_n O} + (1-X_A) \, \mu^0_{B_n O} + n \, RT \, (X_A \, \ln X_A +$$

$$+ (1-X_A) \, \ln (1-X_A)) + n \, X_A \, (1-X_A) \, W_G. \qquad (3.31)$$

From the tangent to the G_{ss} curve (cf. eqn (3.15)) at the
composition of interest we obtain

$$\mu_{A_n O} = \mu^0_{A_n O} + n \, RT \, \ln X_A + n \, W_G \, (1-X_A)^2. \qquad (3.32)$$

The departures of activities from those of ideal
solutions can be conveniently expressed by introducing
<u>activity coefficients</u> γ_i. The activity coefficient is
defined as follows for any n-site solid solution :

$$a_{A_n O} = (X_A \, \gamma_A)^n.$$

By definition, therefore, we have

$$\mu_{A_n O} = \mu^0_{A_n O} + n \; RT \; \ln \; X_A \; \gamma_A$$

$$= \mu^0_{A_n O} + n \; RT \; \ln \; X_A + n \; RT \; \ln \; \gamma_A .$$

(3.33)

In a symmetrical regular solution the values of γ may be obtained from (3.33) and (3.32) :

$$RT \; \ln \; \gamma_A = W_G \; (1-X_A)^2 = W_G \; X_B^2 .$$

(3.34)

Fig. 3.3 shows the molar free energy of a binary $(A,B)_n O$ solid solution at some temperature and pressure at which the G_{ss} curve is inflected into two parts because of the relative magnitudes of G^{xs} and TS_{mix}. At the points X and Y the two parts of the free energy curve have a common tangent. At these points, therefore, the compositions X and Y are in equilibrium since :

$$(\mu_{B_n O})_X = (\mu_{B_n O})_Y$$

$$(\mu_{A_n O})_X = (\mu_{A_n O})_Y$$

(3.35)

This satisfies the condition for equilibrium coexistence of the two phases of composition X and Y. All compositions between the two parts of the G_{ss} curve are unstable with respect to the compositions at X and Y and should therefore react under these conditions to form X plus Y. The latter define the limbs of a miscibility gap in the system.

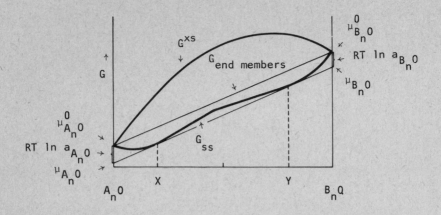

FIG. 3.3. The free energy of an $(A,B)_n O$ symmetrical regular solid solution at some pressure and temperature. Note that solid solutions between X and Y are unstable with respect to the compositions at X and Y which define the limbs of a symmetrical miscibility gap in the system.

Although for a strictly regular solution W_G can only have one value at any pressure, when the regular solution model is applied to real solutions, W_G is found to vary as a function of both pressure and temperature. From the usual form of the expression for G, the interaction parameter may be defined as follows in terms of $W_{H_{1 \, bar}}$, W_S and W_V :

$$W_G = W_{H_{1 \, bar}} - TW_S + (P-1) W_V \qquad (3.36)$$

where $W_{H_{1 \, bar}}$, W_S, and W_V refer to excess (1 bar) enthalpy,

entropy, and volume terms in the mixing properties of the
solid solution. By analogy with (3.30)

$$H^{xs}_{1 \text{ bar}} = n \, X_A \, X_B \, W_{H_{1 \text{ bar}}};$$

$$S^{xs} = n \, X_A \, X_B \, W_S; \quad \text{(for true regular solutions}$$
$$S^{xs} = 0) \qquad\qquad (3.37)$$

$$V^{xs} = n \, X_A \, X_B \, W_V.$$

In general, $H^{xs}_{1 \text{ bar}}$ and S^{xs} have not been determined for
solid solutions of geological interest and must be estimated
from the temperature dependence of W_G using the solution
model. The excess volume, V^{xs}, can be estimated for most
solid solutions from X-ray data. The excess volume is the
difference between the volume of a solid solution derived

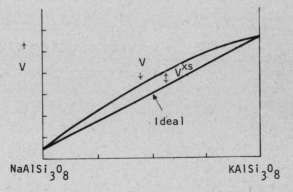

FIG. 3.4. NaAlSi$_3$O$_8$ volumes and V^{xs} of high albite-sanidine
solutions. Note that V^{xs} is a slightly asymmetric function
of composition. Data from Waldbaum and Thompson (1968).

from linear interpolation between the volumes of the end members and the measured volume (Fig. 3.4); in an ideal solution V^{XS} must, by definition, be equal to zero.

3.10 ESTIMATING W_Gs AND ACTIVITY COEFFICIENTS FROM SYMMETRICAL SOLVI

If there is a miscibility gap in the system A_nO-B_nO then it is, in principle, possible to derive activity coefficients and excess parameters for the solid solution $(A,B)_nO$ <u>by assuming</u> that it obeys the regular model. There are several important constraints on this indirect approach to the determination of activity coefficients which must be borne in mind by the reader :

(a)   The miscibility gap should be symmetrical, i.e. the crest of the miscibility surface should be at $X_A = 0.5$ and if at any temperature one limb lies at $X_A = 0.85$, $X_B = 0.15$, the other should also be at $X_b = 0.85$, $X_A = 0.15$.

(b)   The structures of A_nO, B_nO and all solid solutions $(A,B)_nO$ must be the same.

(c)   The entropy of mixing must be ideal, i.e. no ordering of A and B into A-rich and B-rich domains should occur. If ordering does occur, it may be possible to treat <u>some part</u> of the solid-solution series as symmetrical-regular provided the ordering scheme is well characterized.

Given these important restrictions, it is a relatively simple matter to treat a miscibility gap of the type shown in Fig. 3.5 to derive W_G at any temperature and pressure.

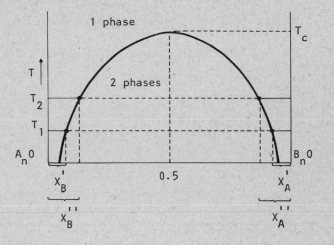

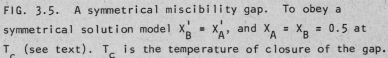

FIG. 3.5. A symmetrical miscibility gap. To obey a symmetrical solution model $X_B' = X_A'$, and $X_A = X_B = 0.5$ at T_c (see text). T_c is the temperature of closure of the gap.

Denoting the two phases coexisting at T_1 by superscripts α(A-rich) and β(B-rich), we have at this temperature the following conditions :

$$\mu^{\alpha}_{B_n0} = \mu^{\beta}_{B_n0}$$

$$\mu^{\alpha}_{A_n0} = \mu^{\beta}_{A_n0}$$

$$n \; RT_1 \; \ln \; a_{B_n O}^{\alpha} = n \; RT_1 \; \ln \; a_{B_n O}^{\beta} \qquad (3.38)$$

$$n \; RT_1 \; \ln \; a_{A_n O}^{\alpha} = n \; RT_1 \; \ln \; a_{A_n O}^{\beta}. \qquad (3.39)$$

Applying the regular solution relationships (3.34), eqn (3.38) becomes

$$n \; RT_1 \; \ln \; X_B^{\alpha} + n \; (1-X_B^{\alpha})^2 \; W_G = n \; RT \; \ln \; X_B^{\beta} + n \; (1-X_B^{\beta})^2 W_G. \qquad (3.40)$$

(An identical relationship for X_A may be derived from (3.39)). Rearranging (3.40) gives

$$RT_1 \; \ln \; \frac{X_B^{\alpha}}{X_B^{\beta}} = W_G \; (1-X_B^{\beta})^2 - W_G \; (1-X_B^{\alpha})^2. \qquad (3.41)$$

Given the measured values of X_B^{α}, X_B^{β}, and temperature, W_G may be readily calculated at the temperature of interest from eqn (3.41). Substitution of this value of W_G into (3.34) enables the calculation of activity coefficients for any $(A,B)_n O$ solid solution at temperature T_1. The same procedure may be used to solve for W_G at all temperatures up to T_c, and therefore to determine W_S and $W_{H_{1 \; bar}}$ from

$$W_G = W_{H_{1 \; bar}} - TW_S + (P-1) \; W_V. \qquad (3.42)$$

The pressure-volume term in eqn (3.42) is usually very small; if required it may be derived by fitting V^{XS} to the measured volumes of the solid solutions (see section 3.9 and eqn (3.37)).

3.11 MULTICOMPONENT SYMMETRIC REGULAR SOLUTIONS

In ternary symmetric solutions, the excess molar free energies of the solution and activity coefficients of components are given by

$$G^{xs} = n X_1 X_2 W_{12} + n X_1 X_3 W_{13} + n X_2 X_3 W_{23} \quad (3.43)$$

$$RT \ln \gamma_1 = X_2^2 W_{12} + X_3^2 W_{13} + X_2 X_3 (W_{12} + W_{13} - W_{23})(3.44)$$

$$RT \ln \gamma_2 = X_1^2 W_{12} + X_3^2 W_{23} + X_1 X_3 (W_{12} + W_{23} - W_{13})(3.45)$$

In these equations W_{ij} refers to the interchange energy for i-j exchanges and X_i to mole fraction.

For four-component symmetric solutions the extension of these equations yields

$$RT \ln \gamma_1 = X_2^2 W_{12} + X_3^2 W_{13} + X_4^2 W_{14} + X_2 X_3 (W_{12} + W_{13} -$$

$$- W_{23}) + X_2 X_4 (W_{12} + W_{14} - W_{24}) + X_3 X_4 (W_{13} +$$

$$+ W_{14} - W_{34}). \quad (3.46)$$

It is rarely necessary to extend the symmetric solution treatment beyond three or four components despite the presence of many minor components in phases of geological interest. Consideration of the equations for G^{xs} and γ_i indicates that unless component j is present in moderately high concentration or W_{ij} is extremely large, interactions between i and j do not make a significant contribution to

γ_i. As an example, let us consider a three-component solution at a temperature of 1000 K with $X_1 = 0.5$, $X_2 = 0.05$, $X_3 = 0.45$, $W_{12} = W_{13} = W_{23} = 2500$ cal (10 460 J). In this three-component solution, γ_1 is given by (3.44) :

$$RT \ln \gamma_1 = 568.75 \text{ cal } (2380 \text{ J})$$

$$\gamma_1 = 1.33.$$

If the contribution from component 2 were ignored, the following result would be obtained :

$$RT \ln \gamma_1 = W_{13} \, X_3^2$$

$$= 506.25 \text{ cal } (2118 \text{ J})$$

$$\gamma_1 = 1.29.$$

The effect of 0.05 mole fraction of component 2 is very small and could in this case be neglected for most calculations. Since most solid solutions only have two, three or four major components, it is necessary to consider only the interactions between these major constituents in order to estimate their activities. (If activities of minor constituents are required, then interactions between the components of interest and major components must of course be considered.)

There is a number of sophisticated solution models which approach more closely the properties of silicates than does the symmetric regular model. The most commonly

used in geological literature is the asymmetric regular
model (Thompson 1967) which has been applied to alkali-
feldspars, (K, Na) micas, olivines, and a number of other
phases. Descriptions of other models may be found in
King (1969) and Saxena (1973).

3.12 ADDITIONAL CAUTIONARY NOTE ON SOLUTION MODELS

If the <u>measured</u> activity-composition relationships for
any particular solid solution fit well to a particular
ideal or non-ideal model over a wide range of temperature
and composition, then it is reasonable to use the model to
extrapolate experimental data to uninvestigated parts of
temperature-composition space. These models may also
be used to derive activity-composition relationships
<u>indirectly</u> using miscibi'' y gaps (section 3.10) or from the
distribution of elements between phases or sites (e.g. Saxena
and Ghose (1971), Ganguly and Kennedy (1974). In the
indirect approach It Is <u>assumed</u> that the thermodynamic
properties of the solid solution obey the relevant equations
without, in most cases, experimental confirmation (except
for V^{xs}). Mixing properties derived from solvi are,
therefore, not necessarily an accurate representation of the
properties of the phase and may be considerably in error,
particularly for low concentrations of one of the components.
Powell (1974) has illustrated the possibility of error in the
indirect approach by fitting the asymmetric regular
(Thompson 1967) and two more complex models, Van Laar and
quasi-chemical, to the sanidine-high albite miscibility gap.
He showed that, although all three models fit the miscibility

gap, the resultant values of γ_i differ considerably and may introduce significant discrepancies between equilibria calculated using the different solution models.

The best values of γ_i to use for calculation are those which have been experimentally measured (see Table 3.1). Indirectly determined γ_i are likely to be less accurate, although, in the absence of experimental data it is probably better to use these than to assume ideality of the solid solution.

3.13   MIXING IN THE FLUID PHASE

Although the preceding discussion of entropies and free energies of mixing has been specifically directed at the properties of solid solutions, much of it also applies to the mixing of fluid and melt components.

Consider one mole of an H_2O-CO_2 fluid containing X_{H_2O} moles of H_2O and X_{CO_2} moles of CO_2 at some pressure P and temperature T. If mixing of H_2O and CO_2 molecules in the fluid is ideal then the molar free energy of the latter is given by

$$G_{fl} = X_{H_2O}\, \mu'_{H_2O} + X_{CO_2}\, \mu'_{CO_2} +$$
$$+ RT\, (X_{H_2O}\, \ln X_{H_2O} + X_{CO_2}\, \ln X_{CO_2})$$

$$(3.47)$$

where μ'_{H_2O} and μ'_{CO_2} are the chemical potentials of pure H_2O and pure CO_2 at P and T (not necessarily the standard state conditions). By analogy with (3.18), the chemical potentials of H_2O and CO_2 in the two-component fluid are

TABLE 3.1

Sources of data on the mixing properties of silicate solid solutions

Phase	Method	Reference
(1) Alkali-feldspar, $KAlSi_3O_8 - NaAlSi_3O_8$	M.G., I.P.	Thompson and Waldbaum (1969)
(2) Plagioclase, $NaAlSi_3O_8 - CaAl_2Si_2O_8$	$E. - 700^{\circ}C$	Orville (1972)
(3) Plagioclase, $NaAlSi_3O_8 - CaAl_2Si_2O_8$	I.P.	Saxena and Ribbe (1972)
(4) Olivine, $Fe_2SiO_4 - Mg_2SiO_4$	$E. - 1150, 1200^{\circ}C$	Nafziger and Muan (1967)
(5) Olivine, $Fe_2SiO_4 - Mg_2SiO_4$	$E. - 1204^{\circ}C$	Kitayama and Katsura (1968)
(6) Olivine, $Fe_2SiO_4 - Mg_2SiO_4$	E. + solution model	Williams (1972)
(7) Olivine, $Fe_2SiO_4 - Mg_2SiO_4$	$E. 1160, 1300^{\circ}C.$	Nafziger (1973)
(8) Olivine, $Fe_2SiO_4 - Mg_2SiO_4$	$E. - 1150^{\circ}C$	Schwerdtfeger and Muan (1966)
(9) Pyroxene, $Fe_2Si_2O_6 - Mg_2Si_2O_6$	E.	(4), (5), (6) above
(10) Orthopyroxene, $Fe_2Si_2O_6 - Mg_2Si_2O_6$	I.P.	Saxena and Ghose (1971)
(11) Orthopyroxene, $Mg_2Si_2O_6 - MgAl_2SiO_6$	I.P.	Wood and Banno (1973)
(12) Mica, $KAl_3Si_3O_{10}(OH)_2 - NaAl_3Si_3O_{10}(OH)_2$	M.G.	Eugster et al. (1972)
(13) Mica, $KFe_3AlSi_3O_{10}(OH)_2 - KMg_3AlSi_3O_{10}(OH)_2$	E.	Wones and Eugster (1965); see also Wones (1971)
(14) Garnet, $Ca_3Al_2Si_3O_{12} - Mg_3Al_2Si_3O_{12}$	$E. 1000 - 1300^{\circ}C$	Hensen et al. (1975)
(15) Garnet, $(Ca, Mg, Fe, Mn)_3Al_2Si_3O_{12}$	Nat	Ganguly and Kennedy (1974)

(16) Magnetite, $Fe_3O_4-Fe_2TiO_4$ M.G. + I.P. } Rumble (1970)

E.   Obtained from experimental measurements

M.G.   Estimated from miscibility gap using solution model

I.P.   Experimentally determined element partition between sites or between phases. Activity-composition relations derived using a solution model.

Nat   Element partition between phases in rocks. Solution model used. Temperature of crystallization assumed.

Order of reliability of data, in general, should be : E > M.G. = I.P. > Nat.

given by

$$\mu_{H_2O} = \mu'_{H_2O} + RT \ln X_{H_2O}$$

$$\mu_{CO_2} = \mu'_{CO_2} + RT \ln X_{CO_2}.$$

(3.48)

If the standard states of H_2O and CO_2 components are taken
to be the pure fluids at the P and T of interest, then we
have

$$\mu'_{H_2O} = \mu^0_{H_2O}$$

$$\mu'_{CO_2} = \mu^0_{CO_2}$$

(3.49)

and

$$a_{H_2O} = X_{H_2O}$$

$$a_{CO_2} = X_{CO_2}.$$

(3.50)

If, on the other hand, the standard states of the fluid
species are taken to be the pure fluids at some pressure P^0
and the temperature of interest, the following relationships
apply for μ'_{H_2O}, μ'_{CO_2} (section 2.9) :

$$\mu_{H_2O}' = \mu_{H_2O}^0 + RT \ln \left(\frac{P\Gamma_{H_2O}}{P^0\Gamma_{H_2O}^0} \right)$$

(3.51)

$$\mu_{CO_2}' = \mu_{CO_2}^0 + RT \ln \left(\frac{P\Gamma_{H_2O}}{P^0\Gamma_{H_2O}^0} \right) .$$

Adding (3.51) and (3.48) yields the following result for chemical potentials and activities of H_2O and CO_2 in the mixed fluid :

$$\mu_{H_2O} = \mu_{H_2O}^0 + RT \ln \left(\frac{P\Gamma_{H_2O}}{P^0\Gamma_{H_2O}^0} \cdot X_{H_2O} \right)$$

(3.52)

$$\mu_{CO_2} = \mu_{CO_2}^0 + RT \ln \left(\frac{P\Gamma_{CO_2}}{P^0\Gamma_{CO_2}^0} \cdot X_{CO_2} \right)$$

$$a_{H_2O} = \left(\frac{P\Gamma_{H_2O}}{P^0\Gamma_{H_2O}^0} \cdot X_{H_2O} \right)$$

(3.53)

$$a_{CO_2} = \left(\frac{P\Gamma_{CO_2}}{P^0\Gamma_{CO_2}^0} \cdot X_{CO_2} \right) .$$

For fluid phases containing any number of components, eqns
(3.51), (3.52), and (3.53) may be applied in analogous ways
to those for solid solutions, provided that mixing of all
components is ideal. In general, of course, mixing of
gaseous species cannot be completely ideal and it is
necessary to introduce activity coefficients γ_i:

$$a_{H_2O} = \left(\frac{P\Gamma_{H_2O}}{P^0\Gamma_{H_2O}}\right) \cdot \quad X_{H_2O} \; \gamma_{H_2O} \cdot \qquad\qquad (3.54)$$

There are, unfortunately, few data available concerning the
activity coefficients of components in fluid phases at
pressures and temperatures of geological interest. There
is some information on H_2O - CO_2 mixing (Greenwood 1973,
Barron 1973) but at the time of writing, the properties of
multicomponent fluids and activities of components other
than H_2O and CO_2 are undetermined. It is common, therefore,
in geological problems, to assume that gaseous species,
although individually imperfect, mix ideally (e.g. Eugster
and Skippen 1967) and that eqns (3.51), (3.52), and (3.53)
apply. This assumption probably does not introduce large
errors in the calculation of equilibria involving fluid
components because such reactions generally involve large
enthalpy and entropy changes. In general, the larger the
enthalpy change of a reaction, the smaller is the error
introduced in calculating equilibrium temperature by an
error in the activity of one of the components involved
(section 4.2).

SUMMARY - CHAPTER 3

1.   Taking the standard state to be the pure phase at the
pressure and temperature of interest, the activity of
component $A_n F_m Si_x O_y$ in the solid solution $(A,B,C)_n$
$(D,E,F)_m Si_x O_y$ is given by :

$$a_{A_n F_m Si_x O_y} = (X_A \gamma_A)^n_{site\ 1} \cdot (X_F \gamma_F)^m_{site\ 2} \qquad (3.55)$$

where

$$(X_A)_{site\ 1} = \left(\frac{A}{A + B + C} \right)_{site\ 1} \quad etc.$$

and γ_A, γ_F refer to activity coefficients for the atoms
A and F on the two sites on which there is mixing.

2.   In an ideal solution, all γ_i are equal to 1.0 and
(3.55) becomes

$$a_{A_n F_m Si_x O_y} = (X_A)^n_{site\ 1} \cdot (X_F)^m_{site\ 2}.$$

3.   For non-ideal solutions the activity coefficient γ_A
is not equal to 1.0. It may, however, be a simple
function of mineral composition. If, for example,
the site of interest can be considered a symmetric
regular solution, then

$$RT \ln \gamma_A = X_B^2 W_G \quad (binary\ A-B\ solution)$$

or

$$RT \ln \gamma_A = X_B^2 W_{AB} + X_C^2 W_{AC} + X_B X_C (W_{AB} + W_{AC} - W_{BC})$$

$$\text{(ternary A-B-C solution)}.$$

W_i may be dependent on temperature and pressure but is a constant, independent of composition, under fixed P,T conditions.

4.   The activity of a component in a fluid phase, taking the standard state to be the pure fluid at the pressure and temperature of interest is given by

$$a_i = X_i \gamma_i. \tag{3.56}$$

If the standard state is taken at the temperature of interest and some pressure P^0, then the following expression is obtained :

$$a_i = \frac{P\Gamma}{P^0\Gamma^0} X_i\gamma_i. \tag{3.57}$$

Activity coefficients γ_i are not known for most mixed fluids of geological interest (except $H_2O\text{-}CO_2$), so the assumption of γ_i equal to 1.0 is generally made.

PROBLEMS - CHAPTER 3

1.   Write down ideal activity-composition relationships (using standard states at the P and T of interest) for end-member components of the following solid solutions :

(a) $(K,Na)(Fe^{2+},Mg)_3AlSi_3O_{10}(OH,F)_2$ biotite

(b) $(Mg,Fe^{2+})_{M2}(Mg,Fe^{2+})_{M1}Si_2O_6$ orthopyroxene.

2.   Given the natural clinopyroxene formula :

Number of ions to 6 oxygens (Deer, Howie, and Zussman 1963, p.114, Analysis No.2)

Si	1.929)	
Al	0.071)	tet

Al	0.049)	
Ti	0.014)	
Fe^{3+}	0.024)	M1
Cr	0.026)	

Mg	0.891)	
Fe^{2+}	0.170)	M1 + M2

Ca	0.78)	
Na	0.024)	M2

calculate the activity of the $CaMgSi_2O_6$ component assuming that Fe^{2+} and Mg^{2+} are divided between M1 and M2 in equal proportions and that there are equal numbers of both positions in the formula unit. Perform the calculation making the two limiting assumptions :

(a) Complete disorder on all types of site.

(b) Mixing of the following charge-balanced 'molecules' takes place.

$Na_{M2}Al^{3+}_{M1}Si_2O_6$; $(Ca,Mg^{2+},Fe^{2+})_{M2}Ti_{M1}Al_2O_6$;

$(Ca,Mg,Fe^{2+})_{M2}(Al,Cr)_{M1}Al^{tet}Si^{tet}O_6$;

$(Ca, Mg^{2+}, Fe^{2+})_{M2} (Mg^{2+}, Fe^{2+})_{M1} Si_2 O_6$.

3.   Consider the equilibrium :

$$3Fe_2Al_4Si_5O_{18} \rightleftharpoons 2Fe_3Al_2Si_3O_{12} + 4Al_2SiO_5 + 5SiO_2.$$
    cordierite   garnet   sillimanite   quartz

The 1 bar enthalpy, entropy, and volume changes for the
reaction involving pure components are as follows :

$\Delta H_{1\ bar} = 38\ 250$ cal (160.04 kJ)

$\Delta S^0 = +\ 26.20$ e.u. (109.6 J K^{-1} mol^{-1})

$\Delta V^0 = -\ 164.2$ $cm^3 = -\ 3.924$ cal bar^{-1}.

Taking standard states of all components to be the
pure phase at the pressure and temperature of interest
and assuming $\Delta C_p = 0$, and ΔV^0 constant, calculate the
following :

(a) Equilibrium pressures at 900 and 1100 K for
all phases pure .

(b) The pressures at 900 and 1100 K under which the
following complex phases would be in equilibrium
(assume ideal solution) :

cordierite $(Mg_{0.6}Fe_{0.4})_2 Al_4Si_5O_{18}$

garnet $(Fe_{0.7}, Mg_{0.2}, Ca_{0.07}, Mn_{0.03}) (Al_{0.98}Fe^{3+}_{0.02})_2$
Si_3O_{12}

sillimanite - pure

quartz - pure

SOLUTIONS TO PROBLEMS - CHAPTER 3

1. (a) $(K,Na)(Fe^{2+},Mg)_3 AlSi_3O_{10}(OH,F)_2$ biotite

Assuming that the mixing on the three distinct types of site is random :

$$a_{KFe_3^{2+} AlSi_3O_{10}(OH)_2}^{biot} = (X_K)_a \ (X_{Fe^{2+}})_b^3 \cdot (X_{OH})_c^2$$

$$a_{NaMg_3^{2+} AlSi_3O_{10}(F)_2} = (X_{Na})_a \ (X_{Mg^{2+}})_b^3 \cdot (X_F)_c^2$$

where the Xs refer to atomic fractions on the three types of site, a, b, c.

Analogous relationships for the components

$NaFe_3^{2+}AlSi_3O_{10}(OH)_2$ $NaMg_3AlSi_3O_{10}(OH)_2$

$KMg_3AlSi_3O_{10}(OH)_2$ $NaFe_3^{2+}AlSi_3O_{10}(F)_2$

$KMg_3AlSi_3O_{10}(F)_2$ and $KFe_3AlSi_3O_{10}(F)_2$

may readily be written.

(b) $(Mg,Fe^{2+})_{M2}(Mg,Fe^{2+})_{M1}Si_2O_6$ orthopyroxene

$$a_{Mg_2Si_2O_6}^{opx} = (X_{Mg})_{M1} \cdot (X_{Mg})_{M2}$$

$$a^{opx}_{Fe_2Si_2O_6} = (X_{Fe})_{M1} \cdot (X_{Fe})_{M2}$$

where

$$(X_{Fe})_{M1} = \left(\frac{Fe^{2+}}{Fe^{2+} + Mg^{2+}}\right)_{M1} \quad etc.$$

Note that activities of the components $Fe_{M1}Mg_{M2}Si_2O_6$ and $Mg_{M1}Fe_{M2}Si_2O_6$ may also be derived but that the latter are, in general, less useful than activities of the end-members because of the lack of standard-state data.

2.   The numbers of each type of site in a clinopyroxene formula unit are as follows : $(M2)_1 (M1)_1 (Tet)_2 O_6$.
After assigning Al, Ti, Fe^{2+}, and Cr to M1 the remaining M1 positions occupied by Fe^{2+} and Mg^{2+} are 0.887 per formula unit. Assigning Ca and Na to M2 leaves 0.196 positions per formula unit occupied by Fe and Mg. Allocating iron and magnesium atoms in equal proportions to the remaining M1 and M2 positions gives :

$$(X_{Mg})_{M1} = 0.735 \quad (X_{Fe})_{M1} = 0.14$$

$$(X_{Mg})_{M2} = 0.156 \quad (X_{Fe})_{M2} = 0.03.$$

(Note that there is a slight deficiency of cations in both M1 and M2 as a result of analytical error.)

(a)   Assuming complete disorder on each type of site and ideal mixing :

$$a_{CaMgSi_2O_6} = (X_{Ca})_{M2} \, (X_{Mg})_{M1} \, (X_{Si})_{tet}^2$$

$$= 0.78 \times 0.735 \times \left(\frac{1.929}{2}\right)^2 = \underline{0.533}.$$

(b)   Assuming mixture as charge-balanced molecules, whenever there is an Mg atom on M1, both tetrahedral positions must be occupied by Si. The latter therefore do not contribute to S_{mix}. The M2 position may be occupied by Ca, Mg, or Fe^{2+} (but not by Na^+). By analogy with eqn (3.25), $a_{CaMgSi_2O_6}$ is obtained from

$$a_{CaMgSi_2O_6} = (X_{Mg})_{M1} \cdot \left(\frac{Ca}{Ca + Mg + Fe}\right)_{M2}$$

$$= 0.735 \times \left(\frac{0.78}{0.976}\right) = \underline{0.587}$$

3.   The reaction :

$$3Fe_2Al_4Si_5O_{18} \rightleftharpoons 2Fe_3Al_2Si_3O_{12} + 4Al_2SiO_5 + 5SiO_2$$
$$\text{cordierite} \qquad\qquad \text{garnet} \qquad \text{sillimanite} \quad \text{quartz}$$

At equilibrium we have

$$\Delta G^0 = - RT \ln \left(\frac{a_{Fe_3Al_2Si_3O_{12}}^{gt\;2} \cdot a_{Al_2SiO_5}^{sill\;4} \cdot a_{SiO_2}^{qz\;5}}{a_{Fe_2Al_4Si_5O_{18}}^{cord\;3}}\right)$$

(a) Taking standard states to be the pure phase at the P and T of interest, the equilibrium conditions for coexistence of pure phases can be obtained from :

$$\Delta G^0 = \Delta H_{1 \ bar} - T\Delta S^0 + (P-1) \ \Delta V^0$$

$$= - RT \ \ln 1 = 0$$

(since all a_i = 1.0 for pure phases).

At 900 K,

$$38250 - 900 \times 26.2 - (P-1) \times 3.9241 = 0$$
$$P = \underline{3738 \ bar}$$

At 1100 K,

$$38250 - 1100 \times 26.2 - (P-1) \times 3.9241 = 0$$
$$P = \underline{2403 \ bar}$$

(b) In the rock, $a^{sill}_{Al_2SiO_5}$ and $a^{qz}_{SiO_2}$ are both equal to 1.0 since both these phases are pure. The impure phases have ideal activities given by

$$a^{cord}_{Fe_2Al_4Si_5O_{18}} = \left(\frac{Fe^{2+}}{Fe^{2+} + Mg^{2+}} \right) = 0.4^2 = 0.16$$

$$a^{gt}_{Fe_3Al_2Si_3O_{12}} = (X_{Fe})^3_c \cdot (X_{Al})^2_{oct}$$

$$= 0.7^3 \cdot 0.98^2 = 0.329.$$

Substituting these values into the equilibrium constant above we have :

$$\Delta G^0 = 38250 - 26.2 \ T - 3.9241 \ (P-1)$$

$$= - \ RT \ \ln \left(\frac{0.329^2}{0.16^3} \right)$$

At 900 K,

$$38250 - 900 \times 26.2 - 3.9241 \times (P-1)$$

$$= - \ 1.987 \times 900 \times 3.2743$$

$$P = \underline{5231 \ bar}$$

At 1100 K

$$38250 - 1100 \times 26.2 - 3.9241 \times (P-1)$$

$$= - \ 1.987 \times 1100 \times 3.2743$$

$$P = \underline{4227 \ bar}.$$

In this way a P-T line for equilibrium coexistence of the complex phases in the rock may be constructed.

4. Geothermometry and geobarometry

4.1 INTRODUCTION

Chapters 1-3 have been devoted to a discussion of the
thermodynamic properties of pure phases and to the
chemical potentials and activities of components in pure
and multicomponent minerals and fluids. The information
given in the earlier chapters may now be used to establish
geothermometers and barometers which enable estimation of
the physical conditions of crystallization of rocks.

The calibration of a mineral geothermometer/
barometer for complex compositions rests on an important
principle which has already been mentioned several times
but may usefully be reiterated here. Let us suppose that
we have standard-state thermodynamic data for the reaction

$$\underset{\text{clinopyroxene}}{2CaMgSi_2O_6} + \underset{\text{olivine}}{Fe_2SiO_4} \rightleftarrows \underset{\text{clinopyroxene}}{2CaFeSi_2O_6} + \underset{\text{olivine}}{Mg_2SiO_4} \quad (4.1)$$

$$\Delta G^0 = 2G^{cpx}_{CaFeSi_2O_6} + G^{ol}_{Mg_2SiO_4} - G^{ol}_{Fe_2SiO_4} - 2G^{cpx}_{CaMgSi_2O_6}$$

$$\text{(at some P,T)}$$

These data enable the calculation of the equilibrium constant K at any pressure and temperature from

$$\Delta G^0 = - RT \ln K = - RT \ln \left(\frac{a^{cpx}_{CaFeSi_2O_6}{}^2 \cdot a^{ol}_{Mg_2SiO_4}}{a^{cpx}_{CaMgSi_2O_6}{}^2 \cdot a^{ol}_{Fe_2SiO_4}} \right). \quad (4.2)$$

If, for example, standard states of all components were taken to be the pure phase at the pressure and temperature of interest, eqn (4.2) would become

$$\Delta H_{1\ bar,T} - T\Delta S^0_T + (P-1)\ \Delta V^0 = - RT \ln K \quad (4.3)$$

(assuming ΔV^0 independent of pressure and temperature). Eqn (4.3) may be used to calculate K at any P and T given 1 bar enthalpy, entropy, and volume changes of reaction (4.1). The form of the equation indicates that if K is constant there is only one equilibrium pressure at each temperature and that fixing K defines an equilibrium curve in P-T space. The standard-state thermodynamic data for the simple reaction (4.1) may be applied to any clinopyroxene-olivine pair crystallized in equilibrium, regardless of their compositional complexities. Given standard-state data and activities of $CaMgSi_2O_6$, Mg_2SiO_4 etc. components in the complex phases, the equilibrium constant for the multicomponent assemblage may be calculated so as to obtain a P-T curve for the assemblage.

Shown in Fig. 4.1 are hypothetical examples of the variation of K with pressure and temperature for equilibria

which would make good geothermometers (Fig. 4.1a) or
geobarometers (Fig. 4.1b).

The factors which control the slopes of the curves of
constant K on the pressure-temperature diagram may be
readily evaluated by differentiating eqn (4.3) with respect
to T at constant pressure and with respect to P at constant
temperature :

$$\left(\frac{\partial \ln K}{\partial T}\right)_P = \frac{-\Delta H_{1\ bar} - (P-1)\ \Delta V^0}{RT^2} \qquad (4.4)$$

$$\left(\frac{\partial \ln K}{\partial P}\right)_T = \frac{\Delta V^0}{RT} \qquad (4.5)$$

If the right-hand side of eqn (4.4) is much larger than the

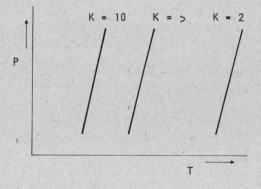

FIG. 4.1(a). Curves of constant K for an equilibrium which
would make a potential geothermometer.

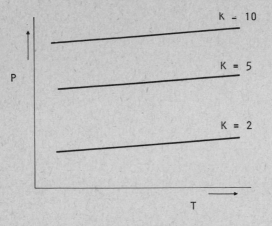

FIG. 4.1(b). A potential geobarometer.

right-hand side of (4.5), K depends much more on temperature than on pressure and the equilibrium would make a
good geothermometer. The opposite condition applies for
a good geobarometer (see Fig. 4.1).

4.2 SELECTION OF USEFUL EQUILIBRIA FOR
   GEOTHERMOMETRY/BAROMETRY

   In any rock containing four or five phases (mineral
and fluid) it is generally possible to write a number of
mass-balanced reactions similar to (4.1) involving
components of the phases present in the rock. Any or all
of these may, in principle, be used to calculate pressure,
temperature, P_{H_2O}, and other intensive variables from
phase compositions and equilibrium constants of the

reactions. If, for example, pressure and temperature
were required, any two equilibria would produce one intersection of two lines of constant K and hence fix pressure and
temperature uniquely. It is obvious that some equilibria
will enable more accurate determination of intensive
variables than others and this section is devoted to
criteria which assist in their selection.

(a) The only useful equilbria are those for which
     reasonably accurate standard-state thermodynamic
     data are either available (e.g. Robie and
     Waldbaum 1968) or may be estimated from simple-
     system experiments (Chapter 7 and section 4.3).

(b) The relationships between activities of the
     relevant components in complex phases and the
     compositions of the phases should be as well defined
     as possible. In practice this means that reactions
     involving major components are generally more
     useful than those involving trace components
     because as concentration increases, γ_i approaches
     1.0.

(c) An equilibrium which is a potential geothermometer
     should have as large a value of $\Delta H_{1 \text{ bar}}$ as possible
     (eqn (4.4)). A good geobarometer should have as
     large a value of ΔV^0 as possible.

The larger the value of $\Delta H_{1 \text{ bar}}$ (positive or negative),
the more rapidly K changes with temperature. This means
that an error in $\Delta H_{1 \text{ bar}}$ or in the activity-composition
relationships for the complex phases will produce a

relatively small error in T if $\Delta H_{1 \text{ bar}}$ is large, and a large error in T if $\Delta H_{1 \text{ bar}}$ is small. Consider, as examples, the two equilibria :

$$KAl_3Si_3O_{10}(OH)_2 + SiO_2 \rightleftarrows KAlSi_3O_8 + Al_2SiO_5 + H_2O \quad (4.6)$$
$$\text{muscovite} \qquad \text{quartz} \quad \text{sanidine} \quad \text{andalusite} \quad \text{fluid}$$

$$\Delta H_{1 \text{ bar}} = 21\ 400 \text{ cal } (89\ 540 \text{ J}) \text{ (Chatterjee and Johannes}$$
$$1974)$$

$$K_6 = \frac{a^{fsp}_{KAlSi_3O_8} \cdot a^{and}_{Al_2SiO_5} \cdot a^{fluid}_{H_2O}}{a^{musc}_{KAl_3Si_3O_{10}} \cdot a^{qz}_{SiO_2}} \cdot$$

$$2KAlSi_2O_6 \rightleftarrows KAlSi_3O_8 + KAlSiO_4 \quad (4.7)$$
$$\text{leucite} \qquad \text{sanidine} \qquad \text{kalsilite}$$

$$\Delta H_{1 \text{ bar}} = -2600 \text{ cal } (-10\ 880 \text{ J}) \text{ (derived from}$$
$$\text{experiments of Scarfe } \underline{et\ al.} \text{ (1967)}$$

$$K_7 = \frac{a^{fsp}_{KAlSi_3O_8} \cdot a^{kals}_{KAlSiO_4}}{a^{lc}_{KAlSi_2O_6}} \cdot$$

Equilibrium curves for (4.6) and (4.7) with all phases pure except feldspar have been calculated using the values of ΔS^0 and ΔV^0_{solids} from Robie and Waldbaum (1968) (Fig. 4.2). The feldspar solid solution was in each case taken to have mole fraction 0.5 of $KAlSi_3O_8$ and 0.5 of $NaAlSi_3O_8$, but two sets of activity-composition relationships for the

$KAlSi_3O_8$ component were used :

(1) $a_{KAlSi_3O_8}^{fsp} = X_{KAlSi_3O_8}^{fsp}$

(2) $a_{KAlSi_3O_8}^{fsp} = X_{KAlSi_3O_8}^{fsp} \, \gamma_{KAlSi_3O_8}$

The activity coefficient $\gamma_{KAlSi_3O_8}$ (about 1.35, but dependent on temperature and pressure) was taken from the asymmetric regular solution treatment of Thompson and Waldbaum (1969).

From Fig. 4.2 it can be seen that the error introduced into the calculated equilibrium temperature by assuming ideal solution is only about $20^{\circ}C$ for reaction (4.6), whereas

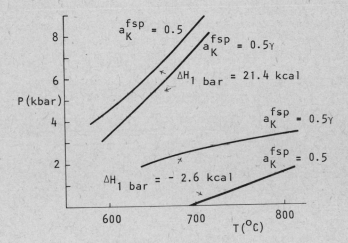

FIG. 4.2. Calculated P-T conditions of equilibrium of $(K_{0.5}Na_{0.5})AlSi_3O_8$ feldspar with the other phases (all pure) involved in reactions (4.6) upper lines) and (4.7) lower lines. See text for discussion.

it is about $150^{\circ}C$ for reaction (4.7). Hence it is possible
to use ideal solution relationships, even for non-ideal
phases, provided $\Delta H_{1\ bar}$ is large. If $\Delta H_{1\ bar}$ is small,
however, it is necessary to know activity-composition
relationships accurately in order to make reliable temperature estimates.

4.3 ALUMINA CONTENT OF ORTHOPYROXENE COEXISTING WITH GARNET

The alumina content of orthopyroxene coexisting with
garnet at any pressure and temperature may be considered
in terms of the simple equilibrium :

$$Mg_2Si_2O_6 \ + \ MgAl_2SiO_6 \ \rightleftharpoons \ Mg_3Al_2Si_3O_{12}.$$

orthopyroxene solid solution   garnet

$$(4.8)$$

(Wood and Banno 1973, Wood 1974). The condition of
equilibrium between any orthopyroxene and any garnet,
regardless of complexity is, of course,

$$\Delta G^0 \ = \ - \ RT \ \ln \left(\frac{a^{gt}_{Mg_3Al_2Si_3O_{12}}}{a^{opx}_{Mg_2Si_2O_6} \cdot a^{opx}_{MgAl_2SiO_6}} \right) . \tag{4.9}$$

Because of the lack of thermodynamic data on
$MgAl_2SiO_6$ orthopyroxene, which has not been synthesized,
it is necessary to estimate $\Delta H^0_{1\ bar}$ and ΔS^0 indirectly.
Wood and Banno (1973) estimated these parameters from

experimental data on the compositions of aluminous
orthopyroxenes in the system $MgO-Al_2O_3-SiO_2$ coexisting with
garnet at known pressures and temperatures. In the system
$MgO-Al_2O_3-SiO_2$ garnet is pure $Mg_3Al_2Si_3O_{12}$ and orthopyroxene
may be regarded as binary $Mg_2Si_2O_6-MgAl_2SiO_6$ solid solution,
so that (4.9) becomes

$$\Delta G^0 = - RT \ln 1 + RT \ln (a^{opx}_{Mg_2Si_2O_6} \cdot a^{opx}_{MgAl_2SiO_6}). \qquad (4.10)$$

Making the simplest activity-composition assumptions for
$Mg_2Si_2O_6-MgAl_2SiO_6$ ($a_i = X_i$), taking standard states to be
the pure phase at the pressure and temperature of interest,
and assuming that ΔV^0 is constant, (4.10) becomes

$$\Delta H_{1\ bar} - T\Delta S^0_T + (P-1)\ \Delta V^0 = RT \ln (X^{opx}_{Mg_2Si_2O_6} \cdot X^{opx}_{MgAl_2SiO_6}) \qquad (4.11)$$

(in the $MgO-Al_2O_3-SiO_2$ system)

The experimentally-determined values of $X^{opx}_{Mg_2Si_2O_6}$ and
$X^{opx}_{MgAl_2SiO_6}$ combined with X-ray determinations of ΔV^0 enable
the estimation of $\Delta H_{1\ bar}$ and ΔS^0_T, provided that two or
more experiments have been performed and the solid solution
is ideal.

There is evidence, however, that eqn (4.11) cannot be
directly apply to Macgregor's (1974) data because
$MgAl_2SiO_6-Mg_2Si_2O_6$ orthopyroxenes have excess volumes of
mixing (V^{xs}) and hence cannot be ideal at high pressures.
This problem was partially circumvented by Wood and Banno

(1973), who assumed ideality at 1 bar and took all non-
ideality to lie in the pressure-volume term of (4.11). It
may readily be shown (eqns (4.40) and (4.43)) that the
activity coefficient of component i at P bar, assuming the
molar and partial molar volumes are independent of pressure,
is given by

$$RT \ln \gamma_i^P = RT \ln \gamma_i^{1 \text{ bar}} + (P-1) (\bar{V}_i - V_i^0). \tag{4.12}$$

Adding an expression of the type (4.12) for each of the
components $Mg_2Si_2O_6$ and $MgAl_2SiO_6$ to eqn (4.11) and taking
$\gamma_i^{1 \text{ bar}}$ to be 1.0 yields

$$\Delta H_{1 \text{ bar}} - T\Delta S_T^0 + (P-1) (V_{Mg_3Al_2Si_3O_{12}}^0 - \bar{V}_{MgAl_2SiO_6}^{opx} -$$

$$\tag{4.13}$$

$$- \bar{V}_{Mg_2Si_2O_6}^{opx}) = RT \ln (X_{Mg_2Si_2O_6}^{opx} \cdot X_{MgAl_2SiO_6}^{opx}).$$

All of Macgregor's data fit eqn (4.13), within
experimental error, yielding $\Delta H_{1 \text{ bar}}^0$ of -7012 cal and
ΔS_T^0 of -3.89 e.u.   These thermodynamic data may now be
used to calculate a P-T line for any orthopyroxene-garnet
assemblage provided that the mixing properties of the more
complex phases present in rocks are known.

In order to apply equilibrium (4.8) to rocks, it may
be possible to assume (Wood and Banno 1973) that complex
garnet and orthopyroxene phases both behave as ideal
solutions.   The relevant activity-composition relationships are (Chapter 3) :

$$a^{gt}_{Mg_3Al_2Si_3O_{12}} = (X_{Mg})^3_c \, (X_{Al})^2_{oct}$$

$$(a^{opx}_{Mg_2Si_2O_6})_{1 \, bar} = (X_{Mg})_{M2} \cdot (X_{Mg})_{M1} \quad\quad (4.14)$$

$$(a^{opx}_{MgAl_2SiO_6})_{1 \, bar} = (X_{Mg})_{M2} \cdot (X_{Al})_{M1} \cdot$$

Substituting the expressions (4.14) into (4.9) gives

$$\Delta H_{1 \, bar} - T\Delta S^0_T + (P-1) \, \Delta V^0 = - \, RT \, \ln \left[\frac{(X^{gt}_{Mg})^3_c \cdot (X^{gt}_{Al})^2_{oct}}{(X^{opx}_{Mg})^2_{M2}(X^{opx}_{Mg})_{M1}(X^{opx}_{Al})_{M1}} \right]$$

$$(4.15)$$

The applicability of eqn (4.15) to complex systems was tested by using experimental data on the systems $FeSiO_3$-$MgSiO_3$-Al_2O_3 and $CaSiO_3$-$FeSiO_3$-$MgSiO_3$-Al_2O_3 and on model rock compositions (Wood 1974). Calculated pressures for experimentally-measured compositions (X^{gt}_{Mg}, $(X^{opx}_{Al})_{M1}$ etc.) of coexisting complex orthopyroxene and garnet are, in most cases, extremely close to the experimental pressures (see Fig. 4.3).

It may surprise the reader, in view of the observed non-ideality of most silicate solid solutions (Chapter 3), that the assumption of ideal solution works so well in this case. The reasons are twofold :

(1) Firstly, both $\Delta H_{1 \, bar}$ and ΔV^0 are quite large for reaction (4.8), thus reducing error due to inadequacies in solution assumptions (section 4.2).

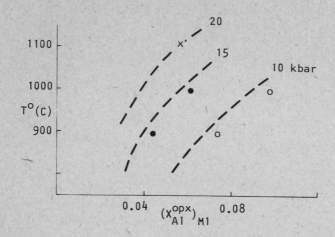

FIG. 4.3. Experimental data on the composition of orthopyroxene coexisting with garnet in the system $CaSiO_3$-
$MgSiO_3$-$FeSiO_3$-Al_2O_3 (Wood 1974) The curves were calculated
using eqn (4.15) and assuming ideal solution in both
phases. x = 20 kbar experiment. o = 10 kbar. ● = 15 kbar.

(2) The second contributing factor may be deduced
     by considering the form of the equilibrium
     constant for reaction (4.8) :

$$K = \frac{a^{gt}_{Mg_3Al_2Si_3O_{12}}}{a^{opx}_{Mg_2Si_2O_8} \cdot a^{opx}_{MgAl_2SiO_6}}$$

     which, may be expressed in terms of site
     occupancies as follows :

$$K_a = \frac{(X_{Mg}^{gt} \; \gamma_{Mg}^{gt})_c^3 \; \cdot \; (X_{Al}^{gt} \; \gamma_{Al}^{gt})_{oct}^2}{(X_{Mg}^{opx} \; \gamma_{Mg}^{opx})_{M2}^2 \; \cdot \; (X_{Mg}^{opx} \; \gamma_{Mg}^{opx})_{M1} (X_{Al}^{opx} \; \gamma_{Al}^{opx})_{M1}} . \quad (4.16)$$

The non-idealities of most silicate solid solutions (see examples in Chapter 3, Table 3.1) are of similar size and magnitude (with γ_i greater than 1) and hence there is a tendency for the activity coefficients in the equilibrium constant (4.16) to cancel out. The result is that (4.16) approximates to the 'ideal' equilibrium constant in eqn (4.15). The tendency for cancelling-out of activity coefficients in the equilibrium constant applies for most reactions involving silicate solid solutions and means that the assumption of ideal solution may often be made with considerable confidence.

4.4 EQUILIBRIA INVOLVING A FLUID PHASE

Equilibria involving a fluid phase are often useful indicators of crystallization temperature. Illustrated in Fig. 4.4 are three experimentally-determined reactions which are relevant to the metamorphism of calcareous (4.17), pelitic (4.18), and ultramafic (4.19) rocks at temperatures in the range 500 to 850°C.

Consider the reactions :

$$\begin{array}{cccccc}
CaCO_3 & + & SiO_2 & \rightleftharpoons & CaSiO_3 & + & CO_2 \\
\text{calcite} & & \text{quartz} & & \text{wollastonite} & & \text{fluid}
\end{array} \qquad (4.17)$$

$$KAl_3Si_3O_{10}(OH)_2 + SiO_2 \rightleftarrows KAlSi_3O_8 + Al_2SiO_5 + H_2O \quad (4.18)$$

muscovite   quartz   sanidine andalusite   fluid

$$4Mg_2SiO_4 + 9Mg_3Si_4O_{10}(OH)_2 \rightleftarrows 5Mg_7Si_8O_{22}(OH)_2 + 4H_2O \quad (4.19)$$

forsterite   talc   anthophyllite   fluid

These all have equilibrium boundaries (and hence curves of constant K) which are more dependent on temperature than on pressure, indicating their potential usefulness as geother-mometers. In addition, all reactions involving a fluid phase have large enthalpy changes and so may be quantitatively

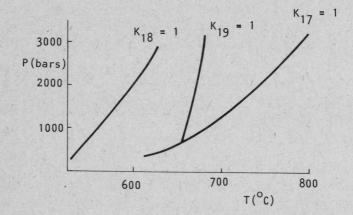

FIG. 4.4. Lines of $K = 1$ (for reactions (4.17), (4.18), (4.19)) involving fluid phases (see text). The standard states of all components have been taken to be the pure phase at P and T of interest. In each case the low temperature assemblage is that on the left-hand side of the reaction as written in the text.

extrapolated to complex systems using approximate (e.g. ideal) activity-composition relations for the mineral phases (section 4.2).

The difficulty which arises when using these equilibria to make temperature estimates is that the composition of the fluid which accompanies metamorphism is not known. The experiments which produced the equilibrium boundaries shown in Fig. 4.4 were performed with a fluid phase of pure H_2O (4.18, 4.19) or pure CO_2 (4.17). In rocks the fluid phase, as well as the mineral phases, need not necessarily be pure. In, for example, metamorphosed dolomitic lime- stones, the common occurrence of hydrated phases such as talc and tremolite indicates that the fluid phase present during metamorphism of these rocks contained H_2O and CO_2 components. It is important, therefore, to determine the dependence of equilibration conditions in such systems on the composition of the fluid. The effect of adding H_2O to a CO_2 fluid phase on the equilibrium conditions for reaction (4.17) may be approximately calculated in the following manner. Let us suppose that the rock contains a mixed H_2O-CO_2 fluid phase and that the total fluid pressure is equal to the total load pressure :

$$P_{fluid} = P_{load}.$$

The total fluid pressure is equal to the sums of the partial pressures of CO_2 and H_2O components :

$$P_{fluid} = P_{CO_2} + P_{H_2O},$$

where

$$P_{CO_2} = P_{fluid} \, X_{CO_2}$$

$$P_{H_2O} = P_{fluid} \, X_{H_2O}$$

and X_{CO_2}, X_{H_2O} refer to mole fractions of H_2O and CO_2 components in the fluid phase. Assuming that H_2O and CO_2 are both imperfect gases but that they mix ideally (e.g. Eugster and Skippen 1967), the activity of the CO_2 component is given by

$$a_{CO_2} = \frac{P_{fluid} \, \Gamma_{CO_2} \, X_{CO_2}}{P^0 \, \Gamma^0_{CO_2}} \qquad (4.20)$$

Taking the standard state of CO_2 to be the hypothetical one of a pure perfect gas at 1 bar (see Appendix 1), the activity becomes

$$a_{CO_2} = P_{fluid} \, \Gamma_{CO_2} \, X_{CO_2}. \qquad (4.21)$$

If the standard-state enthalpy and entropy of reaction (4.17) are known or can be estimated from the equilibrium curve (Chapter 7), then using (4.21) the equilibrium temperature at fixed P and X_{CO_2} can be calculated. Assuming that ΔV^0_{solids} is independent of pressure and temperature, and considering the case of all phases other than the fluid to be pure, the Van't Hoff Isotherm becomes :

$$- RT \ln a_{CO_2} = - RT \ln P \, \Gamma_{CO_2} \, X_{CO_2} = \Delta H^0_{1 \, bar,T} -$$

$$- T\Delta S^0_T + (P-1) \, \Delta V^0_{solids}.$$

(4.22)

Eqn (4.22) may be rearranged and solved for temperature as follows :

$$T = \frac{\Delta H^0_{1 \, bar,T} + (P-1) \, \Delta V^0_{solids}}{\Delta S^0_T - R \ln P \, \Gamma_{CO_2} \, X_{CO_2}}$$

(4.23)

Greenwood (1967) has applied this approach to reaction (4.23) and has confirmed experimentally that the assumption of ideal mixing of H_2O and CO_2 in the fluid is reasonable

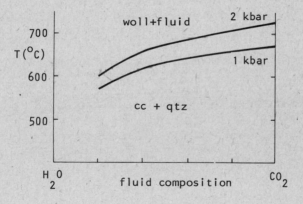

FIG. 4.5. The equilibrium boundary for coexistence of calcite, quartz, wollastonite, and H_2O-CO_2 fluid calculated assuming ideal mixing in the fluid phase (Greenwood 1967).

at 1-2 kbar and $550-700^\circ C$ (see Fig. 4.5). It may be seen
from Fig. 4.5 that, provided the fluid remains richer in
CO_2 than in H_2O, the equilibrium boundary remains within
50° of that for the pure CO_2 fluid. In many cases therefore,
ignorance of the fluid composition and the assumption of P_{CO_2}
equal to P_{fluid} would not produce gross errors in the
calculated equilibrium temperature.

Two additional points should be considered :

(i)   Although accurate determination of fluid phase
       composition in rocks is difficult, some information may,
       in many cases, be obtained from a study of fluid
       inclusions (see, for example, Roedder 1972). It need not
       be necessary, therefore, to assume a specific fluid
       composition.

(ii)   In the preceding discussion it has been implied that
       the presence of a multicomponent fluid complicates the
       application of simple equilibria in geothermometry.
       This need not necessarily be true. If the introduction
       of H_2O in dolomitic marbles produces suitable equilibria
       involving both CO_2 and H_2O components, then the
       occurrence of specific mineral assemblages may enable
       estimates of both temperature and fluid phase composition
       to be made (Skippen 1971, 1974).

The reader interested in metamorphosed pelitic rocks
should refer to the work of Eugster and Skippen (1967) who
have shown that the composition of the fluid phase in
equilibrium with graphite (a common constituent of meta-
pelites) depends greatly on a_{O_2} and temperature. At low

temperatures and low oxygen activities, H_2O is the dominant constituent of the fluid phase coexisting with graphite, whereas at high temperatures and high values of a_{O_2}, CO_2 is dominant. Other species present in amounts which contribute significantly to P_{fluid} are CH_4, CO, and H_2. Given estimates of oxygen activity (see section 4.5 and Rumble 1973), combined calculation of P_{fluid}, temperature, and fluid composition is potentially possible (Jones 1972).

4.5 IRON-TITANIUM OXIDE THERMOMETER/OXYGEN BAROMETER

Many igneous and metamorphic rocks contain the two iron-titanium oxide solid solutions titanomagnetite (Fe_3O_4-Fe_2TiO_4) and hemoilmenite (Fe_2O_3-$FeTiO_3$). The four major components of these two phases may be used to write the following two reactions which enable temperature and oxygen activity to be defined :

$$\underset{\text{titanomagnetite}}{6Fe_2TiO_4} + \underset{\text{fluid}}{O_2} \overset{\rightarrow}{\leftarrow} \underset{\text{titanomagnetite}}{2Fe_3O_4} + \underset{\text{hemoilmenite}}{6FeTiO_3} \quad (4.24)$$

$$\underset{\text{hemoilmenite}}{Fe_2O_3} + \underset{\text{titanomagnetite}}{Fe_2TiO_4} \overset{\rightarrow}{\leftarrow} \underset{\text{titanomagnetite}}{Fe_3O_4} + \underset{\text{hemoilmenite}}{FeTiO_3} \quad (4.25)$$

At equilibrium we have, taking standard states of solid components as the pure phases at P and T and standard state of O_2 as pure O_2 at 1 bar and T,

$$(\Delta H_{1\ bar,T})_{24} - T(\Delta S_T^0)_{24} + \int_1^P (\Delta V_{solids}^0)_{24}\ dP =$$

$$(4.26)$$

$$- RT\ \ln\left(\frac{a_{Fe_3O_4}^{TM^2} \cdot a_{FeTiO_3}^{HI^6}}{a_{Fe_2TiO_4}^{TM^6} \cdot a_{O_2}^{fl}}\right).$$

$$(\Delta H_{1\ bar,T})_{25} - T(\Delta S_T^0)_{25} + \int_1^P (\Delta V_{solids}^0)_{25}\ dP =$$

$$- RT\ \ln\left(\frac{a_{Fe_3O_4}^{TM} \cdot a_{FeTiO_3}^{HI}}{a_{Fe_2O_3}^{HI} \cdot a_{Fe_2TiO_4}^{TM}}\right).\qquad (4.27)$$

Superscripts TM, HI, and fl refer to titanomagnetite, hemoilmenite, and fluid phases respectively. (Note that many authors use the term oxygen fugacity f_{O_2} instead of oxygen activity a_{O_2} in eqns (4.26) and (4.27). To be consistent with the treatment in Chapters 2 and 3, however, we prefer to use fugacity only if the fluid phase is pure. Numerical results are, of course, unaltered by nomen-clature.)   In order to apply eqns (4.26) and (4.27) to the determination of temperature and oxygen activity for any particular rock, a number of thermodynamic and compositional parameters are required : $\Delta H_{1\ bar}^0$, ΔS_T^0, P, ΔV_{solids}^0, activity-composition relations for the solids and the compositions of coexisting multicomponent phases. Instead of measuring these thermodynamic parameters, however, Lindsley (1962, 1963) has experimentally determined the compositions of coexisting titanomagnetite and hemoilmenite phases at known temperatures and oxygen activities. Lindsley's data were subsequently used by Buddington and Lindsley (1964) to calibrate directly the iron-titanium oxide thermometer/oxygen barometer for phases in the system $FeO-Fe_2O_3-TiO_2$ (Fig. 4.6). Two problems associated with the application of the experimental calibration to rocks are immediately obvious :

(a) Lindsley's experiments were all performed at low pressures and P-V terms in eqns (4.26) and (4.27) have been ignored.

(b) In order to apply the data to more complex systems the influence of additional substituents on the activities of components in the two phases should be determined.

It is fortunate in this particular case that lack of these important data does not greatly restrict the application of the curves shown in Fig. 4.6 to rocks. The effect of pressure could not be experimentally detected by Lindsley (1963) and is probably small at pressures up to 10 kbar (Rumble 1970). The concentration of components (MgO, ZnO, V_2O_3 etc.) other than the four major ones are small even in natural solid phases and their influences can be estimated with reasonable confidence. Thus, Buddington and Lindsley's work is of great value for petrologists attempting to determine the magnitude of the intensive variables pertaining

at the time of crystallization of rocks. This is
particularly true in studies of rapidly quenched volcanic
rocks (e.g. Carmichael 1967) in which the compositions
of the oxides present in the lavas correspond to crystallization temperatures and have not changed during cooling. In
more slowly cooled plutonic and metamorphic rocks, iron-
titanium oxides tend to re-equilibrate after crystallization and hence do not always preserve information about the
actual physical conditions of formation.

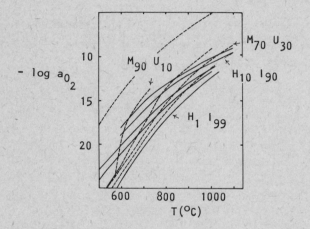

FIG. 4.6. Plot of oxygen activity (fugacity) versus temperature for coexisting iron-titanium oxides. Solid lines show
the compositions of hemoilmenite solid solutions and dashed
lines the compositions of magnetite (M)- ulvospinel (U) solid
solutions. T and a_{O_2} may be determined from the intersection
of appropriate compositional curves.

4.6 Fe^{2+} - Mg EXCHANGE EQUILIBRIA

The distribution of iron and magnesium atoms between coexisting silicate minerals may be described by simple equilibria of the type

$$Mg_2Si_2O_6 \quad + Fe_2SiO_4 \; \rightleftharpoons \; Fe_2Si_2O_6 \quad + Mg_2SiO_4. \quad (4.28)$$
orthopyroxene   olivine   orthopyroxene   olivine

The equilibrium constants for most Fe-Mg exchange reactions are almost independent of pressure because of the small values of ΔV^0 (e.g. Ramberg and De Vore 1951, Banno 1970); this observation has resulted in a number of attempts to calibrate exchange equilibria as geothermometers. Unfortunately, however, most iron-magnesium exchange reactions have small values of $\Delta H^0_{1\ bar}$ as well as ΔV^0 so that the temperature dependence of K (eqn (4.3)) is insufficiently large for accurate temperature estimates to be made. It is probable that the only exchange equilibria which have sufficiently large values of $\Delta H^0_{1\ bar}$ are those involving garnet (clinopyroxene-garnet, cordierite-garnet etc.) and those involving (Mg,Fe^{2+})-spinel.

Irvine (1965) and Jackson (1969) have discussed a thermodynamic approach to the partitioning of Fe^{2+} and Mg between coexisting olivine and spinel. These authors considered the exchange reaction in terms of the simple Fe_2SiO_4 and Mg_2SiO_4 components of olivine and the complex spinel components $Fe^{2+}(Al_x, Cr_y, Fe^{3+}_z)_2O_4$ and $Mg^{2+}(Al_x, Cr_y, Fe^{3+}_z)_2O_4$:

$$\tfrac{1}{2}Fe_2SiO_4 + Mg(Al_xCr_yFe_z^{3+})_2O_4 \rightleftarrows \tfrac{1}{2}Mg_2SiO_4 + Fe^{2+}(Al_xCr_yFe_z^{3+})_2O_4.$$

olivine   spinel   olivine   spinel

(4.29)

In reaction (4.29) x, y, and z are the measured atomic fractions in the octahedral spinel site for the olivine-spinel pair under consideration. Jackson (1969) obtained thermodynamic data for reaction (4.29) and derived a calibration of the empirical distribution coefficient K_D as a geothermometer. The equilibrium constant for reaction (4.29) is

$$K = \frac{a_{Mg_2SiO_4}^{ol\,\tfrac{1}{2}} \cdot a_{Fe(Al_xCr_yFe_z^{3+})_2O_4}^{sp}}{a_{Fe_2SiO_4}^{ol\,\tfrac{1}{2}} \cdot a_{Mg(Al_xCr_yFe_z^{3+})_2O_4}^{sp}} .$$

(4.30)

The empirical distribution coefficient is given by

$$K_D = \frac{X_{Mg}^{ol} \cdot X_{Fe}^{sp}}{X_{Fe}^{ol} \cdot X_{Mg}^{sp}}$$

(4.31)

where

$$X_{Mg}^{ol} = \left(\frac{Mg^{2+}}{Mg^{2+} + Fe^{2+}}\right)_{ol},$$

$$X_{Fe}^{sp} = \left(\frac{Fe^{2+}}{Fe^{2+} + Mg^{2+}}\right)_{sp}.$$

If mixing of Fe^{2+} and Mg^{2+} in both phases is, as assumed by
Jackson, ideal, then K_D is equal to K and temperatures can
be calculated using the standard state thermodynamic data.

Although the olivine-spinel geothermometer gives
apparently reasonable temperatures for chromitite layers in
the Stillwater complex (Jackson 1969) and for some ultramafic intrusions (Loney et al., 1971), Evans and Wright
(1972) found an overestimate of $1000^{o}C$ in the liquidus
temperature of Hawaiian tholeitic lava using this approach.
The reasons for this large discrepancy are the assumption
of ideal Fe^{2+}-Mg mixing in the mineral phases, errors in
the thermodynamic data and difficulties in accurately
determining Fe^{2+}/Fe^{3+} ratios of spinel from microprobe
analyses. It appears, therefore, that this particular
geothermometer is not yet reliable.

4.7 CALCULATION OF EQUILIBRIA INVOLVING SOLID AND
MELT PHASES

Similar methods to those described above may be used to
calculate the conditions under which solid and melt phases
are in equilibrium provided that the thermodynamic
properties of multicomponent melts are known (see Chapter 5).
Although the mixing properties of complex melts are, as
yet, poorly understood, Nicholls and Carmichael (1972) and
Bacon and Carmichael (1973) have described a simple and
potentially powerful approach to the calculation of the
conditions under which magmas form.

Consider a lava quenched rapidly at the earth's surface
(1 bar pressure). During cooling the melt precipitates a
number of mineral phases which, provided conditions
approximate equilibrium, reflect the activities of components
in the silicate liquid. Let us suppose that the minerals
precipitated during quenching of any particular lava are
olivine, clinopyroxene, plagioclase, ilmenite and magnetite.
The oxygen activity and temperature represented by the
quenched assemblage may be calculated from Buddington and
Lindsley's data on coexisting ilmenite and magnetite solid
solutions (section 4.6). We can then write the
equilibrium :

$$
\underset{\text{magnetite}}{2Fe_3O_4} + \underset{\text{glass}}{3SiO_2} \rightleftharpoons \underset{\text{olivine}}{3Fe_2SiO_4} + \underset{\text{vapour}}{O_2} \tag{4.32}
$$

Given the temperature and oxygen activity of the quench,
and olivine and magnetite compositions, the activity of
silica (a_{SiO_2}) in the melt relative to a glass standard state
may be calculated from the equilibrium constant :

$$
K_{32} = \left(\frac{a^{ol}_{Fe_2SiO_4}{}^{3} \cdot a^{fl}_{O_2}}{a^{mt}_{Fe_3O_4}{}^{2} \cdot a^{liq}_{SiO_2}{}^{3}} \right) \tag{4.33}
$$

Bacon and Carmichael (1973) assumed that all solid solutions
are ideal and that the thermodynamic properties of SiO_2 glass
(known) approximate those of SiO_2 liquid (unknown).

Assuming equilibrium between groundmass minerals
and the melt, relationships similar to (4.32) may be written
for a large number of possible melt components. The activity
of alumina ($a_{Al_2O_3}$) is, for example, fixed by the equilibrium:

$$CaMgSi_2O_6 + \tfrac{1}{2}SiO_2 + Al_2O_3 \rightleftarrows CaAl_2Si_2O_8 + \tfrac{1}{2}Mg_2SiO_4.$$

clinopyroxene   glass   liquid   plagioclase   olivine

(4.34)

It is now necessary to consider the activities of
components defined by any solid assemblage with which the
lava may be in equilibrium at high pressures and
temperatures. Let us suppose that one wishes to calculate
the conditions under which the lava would be in equilibrium
with the minerals olivine, orthopyroxene and spinel present
in a spinel lherzolite. Provided solid compositions are
known, the activities of SiO_2 and Al_2O_3 components in any
liquid coexisting with a spinel lherzolite are fixed by the
equilibria

$$Mg_2SiO_4 + SiO_2 \rightleftarrows Mg_2Si_2O_6$$

olivine   glass   orthopyroxene

(4.35)

$$2Mg_2SiO_4 + 2Al_2O_3 \rightleftarrows Mg_2Si_2O_6 + 2MgAl_2O_4.$$

olivine   liquid   orthopyroxene   spinel

(4.36)

From these we have, at any P and T,

$$\ln a_{SiO_2}^{liquid} = \frac{\Delta H_{35}}{RT} - \frac{\Delta S_{35}^0}{R} + (P-1) \frac{\Delta V_{35}^0}{RT} + \qquad (4.37)$$
$$+ \ln a_{Mg_2Si_2O_6}^{opx} - \ln a_{Mg_2SiO_4}^{ol}$$

$$2\ln a_{Al_2O_3}^{liquid} = \frac{\Delta H_{36}^0}{RT} - \frac{\Delta S_{36}^0}{R} + (P-1) \frac{\Delta V_{36}^0}{RT} + \qquad (4.38)$$
$$+ \ln a_{MgAl_2O_4}^{sp^2} \cdot a_{Mg_2Si_2O_6}^{opx} - \ln a_{Mg_2SiO_4}^{ol^2}.$$

(It has been assumed in eqns (4.37) and (4.38) that ΔV^0 is independent of pressure and that solid compositions and component activities are known.)

In order to determine the conditions of equilibrium between melt and solids, it is necessary to know the variation of a_{SiO_2} and $a_{Al_2O_3}$ <u>in the lava of interest</u> as functions of pressure and temperature. The activities of these two components have been determined at 1 bar and the quenching temperature from the compositions of ground-mass minerals. Additional data are, however, required in order to extrapolate these activities to higher pressures and temperatures. The effect of pressure on the activities of components in the melt can be calculated by considering the equation

$$\mu_i = \mu_i^0 + RT \ln a_i. \qquad (4.39)$$

or

$$\mu_i = \mu_i^0 + RT \ln X_i + RT \ln \gamma_i . \tag{4.40}$$

In eqn (4.40), X_i is the mole fraction of i (SiO_2 or Al_2O_3) in the lava, determined from chemical analysis, and γ_i is calculated from the determination of activity at 1 bar and T. Differentiation of (4.40) with respect to pressure at constant temperature and composition (X_i) gives

$$\left(\frac{\partial \mu_i}{\partial P} \right)_{T,X_i} = \left(\frac{\partial \mu_i^0}{\partial P} \right)_{T,X_i} + RT \left(\frac{\partial \ln \gamma_i}{\partial P} \right)_{T,X_i} . \tag{4.41}$$

From eqns (1.39) and (1.40), this expression simplifies to

$$\left(\frac{\partial \ln \gamma_i}{\partial P} \right)_{T,X_i} = \left(\frac{\bar{V}_i - V_i^0}{RT} \right) . \tag{4.42}$$

Nicholls and Carmichael (1972) and Bacon and Carmichael (1973) assumed that molar and partial molar volumes of SiO_2 and Al_2O_3 in the melt are independent of pressure at any given temperature. If correct, this assumption enables (4.42) to be integrated to give the pressure dependence of activity :

$$\ln \gamma_i^{P \text{ bars}} = \ln \gamma_i^{1 \text{ bar}} + (P-1) \left(\frac{\bar{V}_i - V_i^0}{RT} \right) . \tag{4.43}$$

The temperature dependences of γ_{SiO_2} and $\gamma_{Al_2O_3}$ in the melt

are as yet undetermined.   Since the melts are non-ideal
$(a_{SiO_2} \neq X_{SiO_2})$ these authors attempted to extrapolate γ_i to
higher temperatures using the simplest possible model.   They
assumed that the melts behave as regular solutions. If
correct, the regular solution assumption yields the following
relationship for γ_i as a function of temperature (see eqn
(3.34)), with melt composition constant:

$$\ln \gamma_i = \frac{C}{T} . \tag{4.44}$$

C is the sum of all the regular solution parameters for the
multicomponent melt and is determined at 1 bar from a_i,
knowing X_i and the quenching temperature :

$$\ln a_i = \ln X_i + \frac{C}{T} . \tag{4.45}$$

Adding (4.43) to (4.45) gives for the activities of
SiO_2 and Al_2O_3 in the melt

$$\ln a_{SiO_2}^{melt} = \ln X_{SiO_2}^{melt} + \frac{C_{SiO_2}}{T} + (P-1) \left(\frac{\bar{V}_{SiO_2} - V_{SiO_2}^0}{RT} \right) \tag{4.46}$$

$$\ln a_{Al_2O_3}^{melt} = \ln X_{Al_2O_3}^{melt} + \frac{C_{Al_2O_3}}{T} + (P-1) \left(\frac{\bar{V}_{Al_2O_3} - V_{Al_2O_3}^0}{RT} \right) \tag{4.47}$$

Equating the right-hand side of (4.46) with the right-hand
side of (4.37) yields an equation with two unknowns (P and
T) for the conditions under which the melt (lava) and the

lherzolite would be in equilibrium. Following the same
procedure for (4.47) and (4.38) results in two equations
with two unknowns (P and T) which can be solved to obtain
the equilibrium pressure and temperature.

Bacon and Carmichael (1973) used this approach to
determine, for some basanite lavas, the conditions of
equilibrium with the following mineral assemblages which
they contain :

    (a)   spinel lherzolite xenoliths;

    (b)   megacrysts of clinopyroxene and plagioclase;

    (c)   phenocrysts.

The results, illustrated in Fig. 4.7, show the probable
P-T path of the magma, from equilibration with lherzolite
source to final quenching at the earth's surface.

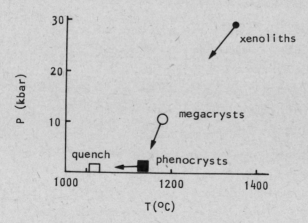

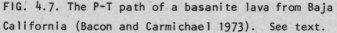

FIG. 4.7. The P-T path of a basanite lava from Baja
California (Bacon and Carmichael 1973). See text.

It is apparent that the method used by these authors
is a potentially powerful tool in petrogenetic studies.
The reader should note, however, that a number of
assumptions have been made :

1. Activity coefficients in melts obey regular
   solution relationships at constant composition.

2. The groundmass phases, precipitated during
   quenching, are in equilibrium with the liquid.

3. The lavas are assumed to rise from source to
   surface without change of composition.

4. The volumes of melt components are independent
   of pressure. Although as yet unverified,
   assumptions 1, 2, and 4 are, fortunately, amenable
   to experimental confirmation.

It is hoped that appropriate experimental tests of the
'Nicholls-Carmichael' approach to petrogenesis will be
carried out in the near future.

4.8 SUMMARY AND CONCLUSIONS

This chapter is not intended to be an exhaustive survey
of available geothermometers and geobarometers. Because
of lack of space a considerable number of suitable
equilibria, particularly those involving miscibility gaps
(diopside-enstatite etc.) and ion exchange have had to be
omitted. The intention has been to show how geothermometers
and barometers may be set up and to illustrate with specific
examples some of the problems (and solutions to these

problems) involved in applying them to rocks. Given the
following data, any simple equilibrium may be applied to
the determination of intensive variables which operated
during rock crystallization : (a) standard state thermodynamic properties of the components involved in the
equilibrium; (b) activity-composition relationships for
multicomponent phases; (c) compositions of the phases
present in the rocks under consideration.

In order that uncertainties in (a), (b), and (c) produce
the smallest possible uncertainties in pressure and
temperature, it is desirable to derive pressures using
equilibria with very large values of ΔV^0 and temperatures
using equilibria with large enthalpy changes :

$$\left(\frac{\partial \ln K}{\partial P}\right)_T \gg \left(\frac{\partial \ln K}{\partial T}\right)_P \tag{4.48}$$

$$\left(\frac{\partial \ln K}{\partial T}\right)_P \gg \left(\frac{\partial \ln K}{\partial P}\right)_T \tag{4.49}$$

Relationships (4.48) and (4.49) should apply, respectively,
for barometers and thermometers.

5. Silicate melts

5.1 INTRODUCTION

The equilibrium conditions of temperature and pressure of
coexisting minerals in solid rocks may be evaluated if the
thermodynamic properties of the mineral and fluid phase
solutions are known (Chapters 1-4). Similar methods may
be used to calculate equilibria involving melt phases
(section 4.7), provided that the mixing properties of
silicate melts are understood, and that the minerals and
melts behave as closed systems.

It should be remembered, however, that during melting
events within the earth, melts may separate from residual
solid phases. Subsequent crystallization or loss of
volatile components may cause further changes in composition
of the evolving magma. Hence, igneous processes rarely take
place under closed-system conditions.

Although the compositions of melts coexisting with
particular mineral assemblages at any given pressure and
temperature may be determined (in each case) by experiment,
the results can be considerably affected by small changes in
composition of the system. It is therefore useful to

develop a thermodynamic treatment of crystal-melt
equilibrium so that the behaviour of natural systems may
be calculated by interpolation from a limited number of
experimental results.

This chapter is devoted to the thermodynamic properties
of silicate melts and to some of the mixing models which
have been proposed to describe their behaviour. Although
the quantitative modelling of multicomponent systems is not
yet possible, our aim is to describe the type of information
already available, and to illustrate some of the areas in
which advances may be made.

5.2 MELTING IN ONE-COMPONENT SYSTEMS

Consider a pure mineral phase in equilibrium with a melt
of the same composition, e.g. pure forsterite at 1 bar,
$1890^{\circ}C$. Since the pure crystal and its melt are in equilibrium, we have

$$\mu^{0}_{Mg_2SiO_4, \; xtal} = \mu^{0}_{Mg_2SiO_4, \; melt} \cdot \qquad (5.1)$$

If we change the temperature or pressure, then equilibrium
between the two phases can only be maintained if the change
in chemical potential is the same in both phases :

$$d\mu^{xtal}_{Mg_2SiO_4} = d\mu^{melt}_{Mg_2SiO_4} \qquad (5.2)$$

Since these changes in μ are given (Chapter 1) by

$$d\mu_i^\alpha = \bar{V}_i^\alpha \, dP - \bar{S}_i^\alpha \, dT, \tag{5.3}$$

then eqn (5.2) can be expressed as

$$\bar{V}_{Mg_2SiO_4}^{ol} \cdot dP - \bar{S}_{Mg_2SiO_4}^{ol} \cdot dT = \bar{V}_{Mg_2SiO_4}^{melt} \, dP$$
$$- \bar{S}_{Mg_2SiO_4}^{melt} \cdot dT \tag{5.4}$$

This may be rearranged to give the Clausius-Clapeyron equation,

$$\frac{dT}{dP} = \frac{\Delta V_m}{\Delta S_m} = \frac{T_m \Delta V_m}{\Delta H_m}, \tag{5.5}$$

where T_m and ΔH_m are the melting point and heat of fusion, and ΔV_m and ΔS_m the volume and entropy of fusion. Since, in general, the volumes and entropies of liquids are greater than those of solids, ΔV_m and ΔS_m are usually both positive, and the melting temperatures increase with increasing pressure (dT/dP positive). Experimental melting curves for forsterite, enstatite, and diopside are shown in Fig. 5.1.

The experimental curves deviate from the straight lines predicted by eqn (5.5) because the compressibilities of the liquids are greater than those of the solids. ΔV_m thus decreases with increasing pressure, causing reduction in the slopes (dT/dP) of the melting curves. This effect can be accommodated using the semi-empirical Simon equation :

SILICATE MELTS

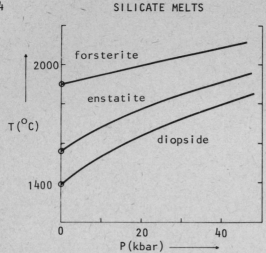

FIG. 5.1. Effects of pressure on the melting temperatures
of silicate minerals.

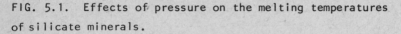

$$P = P_0 + a \left[(T/T_0)^c - 1 \right]. \tag{5.6}$$

where a and c are constant for a given mineral. Values of
a and c for enstatite and diopside are given by Boyd et al.
(1964) :

	T_0(K)	a(kbar)	c
Diopside	1665	23.3	4.64
Enstatite	1830	28.5	5.01

An alternative equation has been found to describe the
melting behaviour of a few pure elements. This has the
form

$$T = T_0 + aP - bP^2 \tag{5.7}$$

where a and b are constants.

Eqn (5.7) predicts a maximum in the melting curve with increasing pressure, beyond which the melting points decrease. Although melting maxima occur in some metallic systems, such behaviour has not, as yet, been observed in any silicate system.

If either the entropy or the volume of fusion is known at any pressure and temperature, than the other may be obtained from (5.5) by differentiating (5.6) :

$$\frac{dP}{dT} = ac(T/T_0)^{c-1} = \frac{\Delta H_m}{T\Delta V_m} \, .$$

5.3 CRYSTAL-LIQUID EQUILIBRIUM - EFFECTS OF OTHER COMPONENTS

In multicomponent systems the chemical potentials in eqn (5.1) must be modified to account for the effects of mixing. In, for example, the case of olivine solid solutions crystallizing from multicomponent melts we have at equilibrium :

$$\mu^0_{Mg_2SiO_4, ol} + RT \ln a^{ol}_{Mg_2SiO_4} = \mu^0_{Mg_2SiO_4, melt} +$$
$$+ RT \ln a^{melt}_{Mg_2SiO_4} . \tag{5.8}$$

Eqn (5.8) can be rearranged to give

$$\ln \frac{a^{melt}_{Mg_2SiO_4}}{a^{ol}_{Mg_2SiO_4}} = - \frac{\mu^0_{Mg_2SiO_4, \, melt}}{RT} + \frac{\mu^0_{Mg_2SiO_4, \, ol}}{RT} \qquad (5.9)$$

Since, however :

$$\frac{\mu^0_i}{RT} = \frac{H^0_i}{RT} - \frac{S^0_i}{R}, \qquad (5.10)$$

differentiation of (5.9) with respect to temperature yields

$$\frac{\partial}{\partial T} \left(\ln \frac{a^{melt}_{Mg_2SiO_4}}{a^{ol}_{Mg_2SiO_4}} \right) = \left(\frac{H^0_{Mg_2SiO_4, \, melt} - H^0_{Mg_2SiO_4, \, ol}}{RT^2} \right) . \qquad (5.11)$$

$\left(H^0_{Mg_2SiO_4, melt} - H^0_{Mg_2SiO_4, ol} \right)$ is the heat of solution of pure Mg_2SiO_4 olivine in a pure Mg_2SiO_4 melt at the temperature T. At the melting point T_0, this is simply the heat of fusion ΔH^0_m or $\Delta H^0_{(T_0)}$.

Eqn (5.11) may be integrated to give the ratio of the activities of Mg_2SiO_4 in the liquid and crystalline phases provided the enthalpy of fusion, ΔH^0_m, is known at the temperature of interest. Generally, ΔH^0_m is known at temperatures near the melting point (T_0) of the pure crystal at 1 bar. Values of ΔH^0_m at other temperatures may be obtained from

$$\int_{T_0}^{T} d\Delta H = \int_{T_0}^{T} \Delta C_p \cdot dT. \qquad (5.12)$$

Substituting into (5.11) and assuming that ΔC_p is constant yields the following expression :

$$\partial \ln \left(\frac{a_{Mg_2SiO_4}^{melt}}{a_{Mg_2SiO_4}^{ol}} \right) = \frac{1}{R} \int_{T_0}^{T} \frac{\Delta H_{(T_0)}^0 + \Delta C_p (T-T_0)}{T^2} \cdot dT. \qquad (5.13)$$

On integration this becomes

$$\ln \left(\frac{a_{Mg_2SiO_4}^{melt}}{a_{Mg_2SiO_4}^{ol}} \right) = \frac{1}{R} \left[(\Delta H_{(T_0)}^0 - \Delta C_p T_0) \left(\frac{1}{T_0} - \frac{1}{T} \right) + \right.$$
$$\left. + \Delta C_p \ln T/T_0 \right]. \qquad (5.14)$$

This is the general equation for the lowering of freezing point of minerals forming solid solutions and crystallizing from multicomponent melts. It is analogous to the Clausius-Clapeyron equation (5.5) discussed above in that both equations describe the conditions required to maintain equilibrium under changing physical or chemical conditions. In eqn (5.5), changes in pressure are offset by changes in temperature (dT/dP) to maintain a pure mineral phase in equilibrium with its melt. In (5.13) changes in bulk composition reduce the activities of the components in the

mineral and melt phases and, to maintain equilibrium at
constant pressure, the temperature must be lowered
$(\partial \ln a_i/\partial T)$. For minerals which crystallize as pure
phases a_i^{xtal} becomes 1.0 in the denominator of eqn (5.14)
(cf. eqn. (5.15)).

This equation may be used to calculate the liquidus and
solidus surfaces if activity-composition relations for both
phases are known, together with the melting point (T_0), heat
of fusion $(\Delta H^0_{(T_0)})$ and ΔC_p (Fig. 5.2).

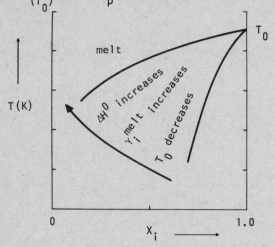

FIG. 5.2. Liquidus curves become less steep as ΔH^0_m increases,
T_0 decreases, and γ_i^{melt} increases.

Alternatively, one may use the large body of equilibrium
data available to obtain activity-composition relationships for
particular silicate melts.

As an example of the use of eqn (5.14), let us consider
the phase diagram for the system Diopside-Albite-Anorthite
which has frequently been used as a simplified basaltic
composition. Although this is not strictly a ternary system
because of solid solution of aluminous components in diopside
(Osborn 1942, Schairer and Yoder 1960), the amount of such
solid solution is small at 1 bar (Kushiro 1973) and so the
system may be considered to be ternary with $a^{di}_{CaMgSi_2O_6} = 1.0$
(Fig. 5.3).

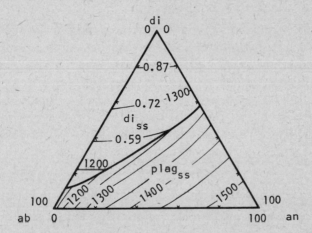

FIG. 5.3. System Diopside-Albite-Anorthite showing the
liquidus fields of clinopyroxene (di_{ss}) and plagioclase.

Diopside melts incongruently (Biggar and O'Hara 1969,
Kushiro 1973) at about 1392°C. Robie and Waldbaum (1968)
give a value of 18 500 cal mol^{-1} for ΔH^0_m and ΔC_p is unknown.

If we assume that ΔC_p is equal to zero, then for the diopside ($CaMgSi_2O_6$) component, eqn (5.14) becomes

$$\ln\left(\frac{a_{di}^{melt}}{1}\right) = \frac{1}{R}\left[18\ 500\left(\frac{1}{1664.7} - \frac{1}{T}\right)\right]. \qquad (5.15)$$

The activity of the diopside component in the melt can thus be estimated at any desired temperature T. For example, along the $1250^{\circ}C$ isotherm $a_{di}^{melt} = 0.59$. Similarly, along the 1300° isotherm $a_{di}^{melt} = 0.72$, and at $1350^{\circ}C$ $a_{di}^{melt} = 0.87$. We could in this way contour the diagram with iso-activity curves and hence obtain activity-composition relationships for components in silicate melts (Fig. 5.3).

It should be noted that these values are, in principle, susceptible to errors in our knowledge of ΔH_m^0 and ΔC_p. For example, if $\Delta C_p = 1.0$, 5.0, or 10.0 cal mol^{-1} K^{-1} at $1250^{\circ}C$ (it is 7.0 for Fe_2SiO_4 according to Kubaschewski, et al. (1967)), then a_{di}^{melt} at $1250^{\circ}C$ becomes 0.596, 0.600, or 0.607 respectively. Clearly in this case ΔC_p need not be known with great accuracy and the assumption that it is equal to zero will not cause significant errors. This is likely to be true whenever ΔH_m^0 is large. Errors of 500 cal in ΔH_m^0 and $\pm 5^{\circ}$ in T_0 have similar effects on the calculated activities.

5.4 ACTIVITY-COMPOSITION RELATIONS FOR SIMPLE MOLTEN SALTS

If the mixing properties of silicate melts were under-

stood it would be possible to use equations like (5.14)
to calculate the effects on observed phase relations of
changes in composition. Studies of the physical properties
of melts help to indicate the type of solution model which
should be adopted.

Electrical conductivity measurements in silicate melts
show that molten silicates are ionic conductors with
relatively free cations and immobile silicate anions
(Bockris et al. 1952).   In a completely ionized molten
salt, the strong Coulombic forces lead to a tendency for
cations to be surrounded by anions and vice versa. The
probability of a cation occupying an anion position or vice
versa is therefore effectively zero. This short-range
ordering makes it possible to carry out a statistical treatment similar to that illustrated for solid solutions in
Chapter 3, by considering a molten salt as a quasi-lattice
with two sets of exclusive 'sites' or energy wells - the
cation and anion matrices.

Consider a mixture of two molten salts M^+B^- and L^+Y^-.
If L^+ and M^+ are sufficiently similar cations and B^- and
Y^- similar anions, then mixing in the cation matrix and in
the anion matrix will approach ideality. The total
configurational entropy of mixing of the two components is
therefore given by

$$S_{mix} = S_{C.M.} + S_{A.M.} \qquad (5.16)$$

where C.M. and A.M. refer to the cation matrix and the anion
matrix respectively.

Hence, from eqn (3.8)

$$S_{mix} = - R\left[(X_L \ln X_L + X_M \ln X_M) + (X_B \ln X_B + X_Y \ln X_Y)\right] \quad (5.17)$$

where X_L, X_M etc. are ion fractions in the respective matrices,

$$\text{i.e. } X_{M^+} = \frac{n_{M^+}}{n_{L^+} + n_{M^+}} \quad \text{etc.} \quad (5.18)$$

Thus for the component AB in an ideal molten salt mixture
we have (cf. eqn (3.28)) :

$$a_{AB} = X_{A^+} \cdot X_{B^-} \cdot \quad (5.19)$$

This quasi-lattice formulation of molten salts was
developed by Temkin (1945) and eqn (5.19) is known as the
Temkin equation.

5.5 MIXING PROPERTIES OF BINARY SILICATE MELTS

The Temkin model of fused salts has been used by a
number of authors to estimate activity composition relationships for silicate melts. Three main types of mixing model
will be considered :

1. Simple ideal (Temkin) mixing (Richardson 1956).

2. Quasi-chemical (Toop and Samis 1962a, b).

3. Polymer models (Masson 1965, Masson et al. 1970;
                    Esin 1973).

Ideal (Temkin) mixing

Richardson (1956) suggested that silicate melts with
the same metal oxide:silica ratio could be considered to mix
ideally according to Temkin's model. This has been found
to be approximately true for a number of binary melts (cf.
Belton et al. 1973). Note that this model does not imply
that, for example, orthosilicates melt according to

$$Mg_2SiO_4 \rightleftarrows 2M^{2+} + SiO_4^{4-} \quad , \qquad\qquad (5.20)$$

but merely that the silicate ions produced do not vary
appreciably between different systems so that there is no
entropy of mixing in the anion matrix. Thus

$$S_{mix} = S_{C.M.} + 0. \qquad\qquad (5.21)$$

This point will be discussed in more detail below.

From (5.21) and (5.19) it can be seen that the
Richardson model may be used to estimate activity-composition
relationships for components in mixtures of silicate melts
having the same metal oxide:silica ratio. For example in
the system Mg_2SiO_4-Fe_2SiO_4, the activities of the Mg_2SiO_4
and Fe_2SiO_4 components in the binary melts are given,
according to this model, by

$$a_{Mg_2SiO_4}^{melt} = X_{Mg, C.M.}^2 \text{ and } a_{Fe_2SiO_4}^{melt} = X_{Fe, C.M.}^2 \qquad (5.22)$$

Since forsterite-fayalite solid solutions are close to
ideal (on sites) (Schwerdtfeger and Muan 1966, Williams

1972), the activities of the components in the solids are approximated by

$$a^{ol}_{Mg_2SiO_4} = X^{M1}_{Mg} \cdot X^{M2}_{Mg} = X^2_{Mg,ol}$$

$$a^{ol}_{Fe_2SiO_4} = X^{M1}_{Fe} \cdot X^{M2}_{Fe} = X^2_{Fe, ol} \cdot$$

(5.23)

These activity-composition relationships (5.22) and (5.23) may be substituted into the depression of freezing point eqn (5.14) to obtain the expressions

$$2 \ln \left(\frac{X^{melt}_{Mg}}{X^{ol}_{Mg}} \right) = \frac{1}{R} \int_{2164}^{T} \frac{\Delta H^0_{Fo}(T_0) + \int_{2164}^{T} \Delta C_p \cdot dT}{T^2} \cdot dT \quad (5.24)$$

and

$$2 \ln \left(\frac{X^{melt}_{Fe}}{X^{ol}_{Fe}} \right) = \frac{1}{R} \int_{1479.5}^{T} \frac{\Delta H^0_{Fa}(T_0) + \int_{1479.5}^{T} \Delta C_p \cdot dT}{T^2} \cdot dT \quad (5.25)$$

Using the values

$$\Delta H^0_{Fo} = 34\ 240\ cal\ mol^{-1} \text{ (estimated from phase diagram)}$$

$$\Delta H^0_{Fa} = 25\ 010\ cal\ mol^{-1} \text{ (estimated from phase diagram)}$$

$$\Delta C_{p(Fo)} = \Delta C_{p(Fa)} = 7.45\ cal\ mol^{-1}\ deg^{-1} \text{ (Orr 1953)},$$

eqns (5.24) and (5.25) may be used to calculate the ratios

$(X_{Mg}^{melt}/X_{Mg}^{ol})$ and $(X_{Fe}^{melt}/X_{Fe}^{ol})$ required to maintain crystal-melt equilibrium at various temperatures (T) as shown in Table 5.1.

TABLE 5.1

Temp (K)	1479.5	1548	1600	1683	1738	1768
Fe^{melt}/Fe^{ol} (= a) :	1.0	1.21	1.39	1.70	1.93	2.06
Mg^{melt}/Mg^{ol} (= b) :	-	0.23	0.27	0.34	0.40	0.43

These values may be used to solve uniquely for the liquidus and solidus curves from the two simultaneous equations :

$$X_{Fe}^{melt} = a\ X_{Fe}^{ol} \tag{5.26}$$

$$(1-X_{Fe}^{melt}) = b\ (1-X_{Fe}^{ol}) \tag{5.27}$$

where a and b are values of the ratios shown in Table 5.1. Thus

$$X_{Fe}^{melt} = \frac{a\ (1-b)}{a-b}\ ;\quad X_{Fe}^{ol} = \frac{(1-b)}{(a-b)}\ . \tag{5.28}$$

The compositions of coexisting olivines and melts calculated from (5.28) are compared with Bowen and Schairer's (1935) experimental data for the same temperatures in Fig. 5.4.

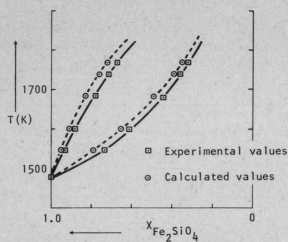

FIG. 5.4. Comparison of experimentally determined Mg_2SiO_4-Fe_2SiO_4 phase diagram with values calculated using Richardson's model.

Despite the uncertainties in the input values of ΔH^0_{Fo}, ΔH^0_{Fa}, ΔC_p and $T_{0(Fo)}$, the experimental and calculated phase diagrams agree quite well.

It is worth stressing at this point that calculations of this type are not a substitute for direct experimental observation. There is little point in recalculating phase diagrams which have already been well investigated other than to check on the accuracy of the proposed theoretical model.

The value of the thermodynamic approach lies in its ability to predict the effects of varying the conditions (pressure, a_{H_2O}, a_{CO_2}, a_{O_2} etc.) from those of the specific experimental runs in question. For example, the addition of

H_2O to the system Diopside-Forsterite-Silica causes the
forsterite field to expand with increasing H_2O-content
(Kushiro 1969, 1972). Increasing total pressure or P_{CO_2}
has the opposite effect (Kushiro 1969, Eggler 1973,
1976, Mysen and Boettcher 1975) and changes in the
contents of other added oxide components may also have
significant effects on melting relations determined in a
simpler system, as is shown in Fig. 5.5.

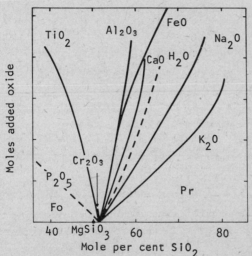

FIG. 5.5. Effects of added oxide components on the
position of the forsterite-protoenstatite cotectic in the
systems $MgO-SiO_2$-other oxide after Kushiro 1975.

It is clear that melting and crystallization behaviour
determined from experimental work in simple systems may
have to be modified if other components are present in the
natural system in question. Accurate predictions of the

likely effects of varying conditions are of great value
to those who do not have ready access to experimental
facilities, and in helping the experimentalist to decide the
conditions under which his experiments should be performed.

Although the ideal Temkin mixing model was used above
to estimate the activities of components in silicate melts
with fixed silica contents, in most geological cases
minerals crystallize from melts of quite different metal
oxide:silica ratios (cf. the forsterite field in the
system $MgO-SiO_2$).   In these cases the simple Richardson
model cannot be applied to estimate activities in the melts.

Two more complex mixing models have been applied to
silicate melts with varying SiO_2 contents and these will be
briefly reviewed below.

Silicate melts with varying silica contents. The Toop and Samis (quasi-chemical) model

Consider the mixing of a basic oxide (e.g. CaO) with a
silicate melt of higher silica content (e.g. SiO_2).   As
more basic oxide is added, the three dimensionally cross-
linked silica network is successively broken down into
lower polymers. This process may be represented
schematically by :

$$O^{2-} \quad + - \overset{|}{\underset{|}{Si}} - O - \overset{|}{\underset{|}{Si}} - \rightleftharpoons 2 - \overset{|}{\underset{|}{Si}} - O^-.$$

Free oxide   Silicate melt   (5.29)

Following Fincham and Richardson (1954), Toop and
Samis (1962a, b) made the simplifying assumption that in

reaction (5.29) the 'reactivity' of an oxo-bridge in a
silicate polymer is independent of the molecular size of the
particular polyanion involved. This assumption has been
found to apply to many organic polymer systems and is the
result of the poor transmission of bonding information along
polymeric molecules.

Letting the reactivity of oxo-bridges be independent of
molecular weight, the reaction (5.29) may be expressed as

$$O^{2-} + O^0 \rightleftharpoons 2 \, O^- \tag{5.30}$$

where

O^{2-} = free oxide ion,

O^0 = bridging oxygen atom,

O^- = singly-bound oxygen.

Note that (5.30) is not a reaction among discrete molecules,
but is written in terms of three quasi-chemical oxygen
species.

The use of quasi-chemical species to describe the mixing
properties of silicate melts can be seen to be a refinement
of the regular solution model discussed in Chapter 3. In a
perfect regular solution, there is a heat of mixing but
S_{mix} is considered to be ideal. This is unlikely to be true
if large heats of mixing are observed as in the case of
silicate melts with different silica contents because the
heats of mixing correspond to attractions or repulsions
among the mixing components.

A more refined treatment incorporates these effects by
taking account of reactions between the components so as to
produce reacted species. This refined regular solution model
expresses non-ideal mixing behaviour (e.g. of A and B) in
terms of the ideal mixing of these quasi-chemical (reacted)
species. This is rather similar to a change of standard
state (cf. Chapter 2) with the inclusion of the RT ln a_i
term in μ_i^0 (compare the behaviour of $\int V\,dP$ in the 1 bar, T
and P,T standard states). In general, it is useful to note
that non-ideal behaviour can often be considerably
simplified through a judicious choice of components.

In order to apply the quasi-chemical mixing model (5.30)
to estimate activities in silicate melts, Toop and Samis
assumed that :

1. The Temkin model holds in binary silicate melts
   $MO\text{-}SiO_2$. All the free energy of mixing is therefore
   anionic (eqn (5.30)) since $X_{M^{2+}}^{C.M.} = 1.0$.

2. Metal oxide components, MO, are 100 percent dissociated.

3. The reactivities of oxo-bridges are independent of
   molecular size and configuration.

Making these assumptions, Toop and Samis considered the
reverse of (5.30) :

$$2\ O^- \rightleftarrows O^0 + O^{2-}. \tag{5.31}$$

If the quasi-chemical species are considered to mix
ideally, the equilibrium condition for (5.31) becomes :

$$K = \frac{a_O^0 \cdot a_O^{2-}}{a_O^{2-}} = \frac{X_O^0 \cdot X_O^{2-}}{X_O^{2-}} . \qquad (5.32)$$

The free energy change for the reaction of a basic oxide MO with SiO_2 (5.30) is then given from (5.32) by

$$\Delta G^0 = - RT \ln (1/K) = RT \ln K. \qquad (5.33)$$

Since one mole of O^{2-} reacts with O^0 to produce 2 moles of O^-, the free energy of mixing per mole of silicate melt formed is given by :

$$G_{mix} = RT \ln K \ (\tfrac{1}{2} n_O^-) \qquad (5.34)$$

n_O^- is the number of moles of O^- species in the melt. It is calculated by assuming that 1 mole of MO produces 1 mole of O^{2-} ion, 1 mole of SiO_2 contributes 2 moles of O^0 and that O^-, O^{2-} and O^0 are related by (5.32). n_O^- is then given by the quadratic equation :

$$n_O^2 - (4K - 1) + n_O^- (2 + 2X_{SiO_2}) + 8X_{SiO_2}(X_{SiO_2} - 1) = 0. \qquad (5.35)$$

Curves of G_{mix} can be calculated at different bulk compositions (X_{SiO_2}) for given values of K from eqns (5.35) and (5.34). Examples of these curves are shown in Fig. 5.6.

It can be seen that the calculated curves agree closely with the experimental values if appropriate values of K are

SILICATE MELTS

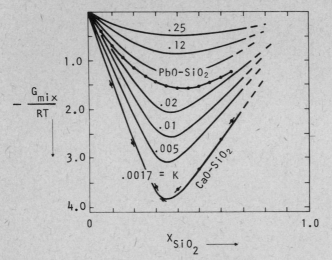

FIG. 5.6. Comparison of values of G_{mix} calculated using the Toop and Samis model with experimental data for $CaO-SiO_2$ and $PbO-SiO_2$ melts.

chosen for each system. The equilibrium constants K in (5.32) are therefore characteristic of the cations present.

Some applications of the Toop and Samis model

(i) <u>Uptake of H_2O, CO_2, sulphur by silicate melts</u>. It is well known that the solubility of water in silicate melts increases with the square root of water fugacity (Fig. 5.7).

The linear dependence of water solubility on the square root of f_{H_2O} may be explained in the following manner using the Toop and Samis model.

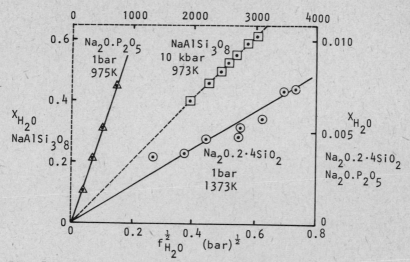

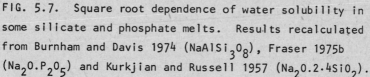

FIG. 5.7. Square root dependence of water solubility in some silicate and phosphate melts. Results recalculated from Burnham and Davis 1974 ($NaAlSi_3O_8$), Fraser 1975b ($Na_2O \cdot P_2O_5$) and Kurkjian and Russell 1957 ($Na_2O \cdot 4SiO_2$).

Since the viscosities of silicate melts are decreased markedly by the absorption of H_2O, it is likely that, at least in acidic melts, water is absorbed in the melt by reaction with oxo-bridges causing depolymerization. If the reactivities of the oxygen species are independent of molecular size, then the absorption of H_2O may be expressed by :

$$H_2O \ (vap) \ + \ O^0 \ \rightleftarrows \ 2 \ OH, \tag{5.36}$$

where OH represents an OH group attached to a silicate polymer, i.e. $- \overset{|}{\underset{|}{Si}} - OH$.

If the quasi-chemical oxygen species mix ideally then the equilibrium constant for (5.36) is

$$K = \frac{X_{OH}^2}{f_{H_2O} \cdot X_{O^0}} = \frac{(2X_{H_2O}^{melt})^2}{f_{H_2O} \cdot X_{O^0}} \quad . \tag{5.37}$$

Thus the equilibrium water content $(X_{H_2O}^{melt})$ increases with $(f_{H_2O})^{\frac{1}{2}}$. The results of Burnham and Davies (1974) on albite-H_2O may be expressed in this form by assuming that one mole of $NaAlSi_3O_8$ melt contains eight moles of O^0. Values of ln K at different pressures calculated from eqn (5.37) are plotted against 1/T in Fig. 5.8.

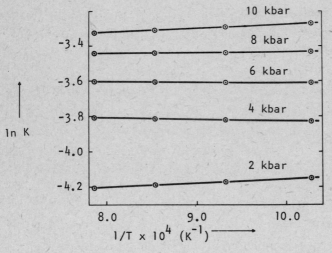

FIG. 5.8. Solubility of H_2O in $NaAlSi_3O_8$ melts expressed using the Toop and Samis model (after Fraser 1975b).

The solubility of CO_2 in silicate melts may be expressed in a similar manner :

$$CO_2 \text{ (vap)} + 0^{2-} \rightleftarrows CO_3^{2-} . \qquad (5.38)$$

The solubility of CO_2 as carbonate in silicate melts should therefore increase with increasing basicity of the melt. This has been demonstrated experimentally by a number of workers (e.g. Mysen et al. 1976).

The solubility of sulphur in some silicate melts was determined by Fincham and Richardson (1954), who found that two mechanisms were involved depending on the oxygen activity :

$$\text{(a)} \quad \tfrac{1}{2} S_2 \text{ (vap)} + 0^{2-} \text{ (melt)} \rightleftarrows \tfrac{1}{2} O_2 \text{ (vap)} + S^{2-} \text{ (melt)} \quad (5.39)$$

$$\text{(b)} \quad \tfrac{1}{2} S_2 \text{ (vap)} + 0^{2-} \text{ (melt)} + {}^3/_2 \, O_2 \text{(vap)} \rightleftarrows SO_4^{2-} \text{ (melt)}. \quad (5.40)$$

Thus at low a_{O_2} values, sulphur dissolves in silicate melts primarily as sulphide, S^{2-}, and the equilibrium sulphur content decreases with increasing a_{O_2}. However, with further increases in oxygen activity, reaction (5.40) becomes important and the equilibrium sulphur contents (as SO_4^{2-}) of the melts then increase with increasing a_{O_2}. In both cases basic melts have higher sulphur solubilities than acidic melts because of the dependence of both reactions on oxide ion activity.

(ii) Effects of melt composition on oxidation state. In
addition to the effects of melt composition on the
solubilities of gases, it has been found that the
composition of the melt may have a strong influence on
the oxidation states of altervalent components in the melt
at constant a_{O_2}. This has been observed for many systems
including Fe^{3+}/Fe^{2+}, Mn^{3+}/Mn^{2+}, Eu^{3+}/Eu^{2+}, and Cr^{6+}/Cr^{3+}
(e.g. references in Fraser 1975a). Since the oxidation
ratio may vary by an order of magnitude or more with
changing melt composition, distribution reactions (see
Chapter 6) involving elements which can be present in
different redox states may be strongly dependent on gross
melt composition as well as oxygen activity. Suggested
use of such reactions as monitors of a_{O_2} during magmatic
processes (Drake 1975) should therefore be treated with
caution.

The effects of melt composition on oxidation state
cannot be accounted for by using simple reactions analogous
to those written above to express gas solubility. For
example, the oxidation state of europium in silicate melts
(Morris and Haskin 1974) is not described by the equation

$$4 \, Eu^{3+} \; + \; 2 \, O^{2-} \; \rightleftarrows \; 4 \, Eu^{2+} \; + \; O_2 \, . \tag{5.41}$$
$$\text{(melt)} \quad \text{(melt)} \quad \text{(melt)} \quad \text{(vap)}$$

This reaction predicts that Eu^{2+} should be stabilized
in more basic melts (higher $a_{O^{2-}}$). However, the opposite
is observed experimentally.

The reason for this anomaly may lie in the different acid-base properties of Eu_2O_3 and EuO (Fraser 1975a), as can be seen by writing (5.41) in the form

$$Eu_2O_3 \rightleftharpoons 2EuO + \tfrac{1}{2}O_2.$$
(melt)   (melt)

(5.42)

The activities of each oxide in (5.42) will be determined by the reactions of Eu_2O_3 and EuO with the silicate melt in question.

EuO is likely to be a basic oxide like CaO, dissociating to give Eu^{2+} and O^{2-} :

$$EuO \rightleftharpoons Eu^{2+} + O^{2-}.$$

(5.43)

In the case of Eu_2O_3, however, it was found necessary to express its behaviour as an amphoteric oxide :

$$Eu_2O_3 \rightleftharpoons 2Eu^{3+} + 3O^{2-} \quad \text{'Basic' reaction, } K_b$$

(5.44)

$$Eu_2O_3 + O^{2-} \rightleftharpoons 2EuO_2^-$$

or possibly $\left.\rule{0pt}{36pt}\right\}$ 'Acidic' reaction, K_a.

$$Eu_2O_3 + 5O^{2-} \rightleftharpoons 2Eu_4^{5-}$$

(5.45)

Proceeding in this way, it was shown that the ratio $Eu^{III}/^{II}$ can be expressed by an equation of the form

$$\frac{Eu^{III}}{Eu^{II}} = \frac{a_{O_2}^{\frac{1}{4}}}{Const} \left[K_a^{\frac{1}{2}} (a_{O^{2-}})^{3/2} \cdot \frac{(1-X_{MO})}{X_{MO} + \frac{1}{4}(1-X_{MO})} + \frac{K_b^{\frac{1}{2}}}{(a_{O^{2-}})^{\frac{1}{2}}} \right].$$

(5.46)

This equation predicts that the ratio Eu^{III}/Eu^{II} is proportional to $a_{O_2}^{\frac{1}{4}}$ at constant composition, as is observed experimentally both for Eu^{III}/Eu^{II} and Fe^{III}/Fe^{II} (Paul and Douglas 1965). If Eu_2O_3 behaved as a basic oxide, dissociating to give Eu^{3+} and O^{2-} ions, then in eqn (5.46) $K_a \rightarrow 0$ and the ratio Eu^{III}/Eu^{II} would decrease with increasing oxide activity (cf. eqn (5.41). However, if Eu_2O_3 is _amphoteric_ and the acidic reaction (5.45) predominates, then the expression involving K_a becomes important so that the Eu^{III}/Eu^{II} ratio _increases_ with increasing basicity as is observed experimentally.

Masson's polymer models

A slightly different approach has been developed by Masson and his co-workers (Masson 1965, Masson _et al._ 1970) and also by Esin (1973). In the simpler version (Masson 1965), the polymerization of unbranched silicate anions in the melt is considered:

$$SiO_4^{4-} + SiO_4^{4-} \rightleftarrows Si_2O_7^{6-} + O^{2-} \qquad (K_1) \qquad (5.47)$$

$$SiO_4^{4-} + Si_2O_7^{6-} \rightleftarrows Si_3O_{10}^{8-} + O^{2-} \qquad (K_2) \qquad (5.48)$$

$$\vdots \qquad\qquad \vdots \qquad\qquad\qquad\qquad \vdots$$

$$SiO_4^{4-} + Si_nO_{3n+1}^{(2n+2)} \rightleftarrows Si_{n+1}O_{3n+4}^{(2n+4)} + O^{2-} .(K_n) \qquad (5.49)$$

Making a similar assumption to that of Toop and Samis, Masson proposed that $K_1 = K_2 = \cdots = K_n = K$

where

$$K_1 = \frac{X_{Si_2O_7} \cdot X_{O^{2-}}}{X_{SiO_4} \cdot X_{SiO_4}} \quad \cdot \quad \frac{\gamma_{Si_2O_7} \cdot \gamma_{O^{2-}}}{\gamma_{SiO_4} \cdot \gamma_{SiO_4}} \quad \text{etc.} \qquad (5.50)$$

Masson found that if the activity coefficients in (5.50) were omitted, the infinite number of polymerization reactions could be simplified to yield :

$$X_{SiO_4} = \frac{1 - X_{O^{2-}}}{1 + K \left(\frac{1}{X_{O^{2-}}} - 1\right)} \cdot \qquad (5.51)$$

The ion fractions of other polymers can then be obtained from eqns (5.47) - (5.49). By summing up the number of moles of SiO_2 in all the polymers, an expression can be obtained for X_{SiO_2} in terms of $X_{0^{2-}}$ and K :

$$X_{SiO_2} = \cfrac{1}{\left[3-K + \cfrac{X_{0^{2-}}}{1-X_{0^{2-}}} + \cfrac{K\ (K-1)}{\cfrac{X_{0^{2-}}}{1-X_{0^{2-}}} + K} \right]} \qquad (5.52)$$

This equation expresses the ion fraction of free oxide $(X_{0^{2-}})$ as a function of mole fraction SiO_2 in binary silicate melts.

Masson finally used the Temkin eqn (5.19) in the form

$$a_{MO} = X_{M^{2+}} \cdot X_{0^{2-}}$$

to calculate activity-composition relationships for oxide components in silicate melts which were in good agreement with experimental values (Fig. 5.9).

A more general polymerization model has since been developed (Whiteway et al. 1970) to allow the possibility of branched silicate chains. In this case, (5.52) becomes

$$\frac{1}{X_{SiO_2}} = 2 + \frac{1}{1-X_{0^{2-}}} - \frac{3}{1 + X_{0^{2-}}\ (\frac{3}{K} -1)} \qquad (5.53)$$

It should be noted that these expressions do not include activity coefficients. The good agreement between theoretical and calculated oxide activities implies either that the different polymers mix ideally, which is unlikely, or that

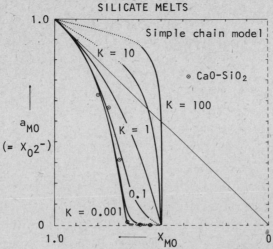

FIG. 5.9. Activity-composition relationships for oxide
components in silicate melts. Curves calculated from eqn
(5.52) using different values of K. Note that curves are
assumed to pass through $X_{O^{2-}} = 1.0$ at the pure MO composition.

the γs in (5.50) cancel out. To cancel out entirely, the
activity coefficients must lie in a geometrical series (if
they are not all equal) with

$$\frac{\gamma_{O^{2-}}}{\gamma_{SiO_4}} = \frac{\gamma_{SiO_4}}{\gamma_{Si_2O_7}} = \cdots = \frac{\gamma_{Si_nO_{3n+1}}}{\gamma_{Si_{n+1}O_{3n+4}}} \qquad \text{etc.}$$

This implies that the tetrahedral segments of the silicate
polyanions mix ideally, a result which may be useful in
developing activity-composition relationships for silicate
components in complex melts.

5.6 EXTENSION TO MULTICOMPONENT SYSTEMS

Natural rock melts seldom even approach the simplicity of the binary systems treated above. In order to treat igneous systems effectively it will be necessary to consider the mixing properties of oxides which are clearly amphoteric (e.g. Al_2O_3 and Fe_2O_3). The development of an acid-base treatment as outlined above may be useful in this context since, as was shown, the mixing properties of amphoteric oxides can be split into end-member acidic and basic reactions characterized by the constants K_a and K_b. In principle, oxides can be divided up on this basis to form a spectrum ranging from acidic to basic in properties :

$$P_2O_5,\ TiO_2\ \ldots\ Al_2O_3,\ Fe_2O_3\ \ldots\ CaO,\ K_2O.$$

acidic	amphoteric	basic
$K_b \to 0$	$K_a,\ K_b$	$K_a \to 0$

However, it should be noted that the Masson polymer models have not yet been applied outside binaries. One reason for this may lie in the assumption that the oxides other than SiO_2 are completely dissociated. Any differences in the degree of dissociation (e.g. between CaO and MgO) will be reflected in changes in the standard-state value assumed for $X^0_{O^{2-}}$ in the pure liquid oxides (Fraser 1975a).

5.7 BURNHAM'S MIXING MODEL

An alternative means of examining the mixing properties of silicate melts has been suggested by Burnham (1975).

This was developed from a study of the solubility of H_2O in $NaAlSi_3O_8$ melts at different pressures and temperatures (Burnham and Davis 1974). The solubility measurements showed that at water concentrations in the melt of up to at least 50 mol percent, the solubility increases with the square root of water fugacity (see Fig. 5.6).

As discussed above, the linear dependence of water solubility on the square root of f_{H_2O} implies that H_2O dissociates in the melt to form two independent species.

It should be noted that any proposed reaction in which two moles of independent species are produced in the melt per mole of H_2O absorbed will be able to account for the observed square root dependence of the solubility on f_{H_2O}.

To interpret these results Burnham (1975) has proposed that the following equilibrium describes the behaviour of water in albite melts,

$$H_2O + O^{2-} + Na^+ \rightleftharpoons OH^- + ONa^- + H^+. \qquad (5.54)$$

This reaction appears unlikely to the authors since it involves the existence of free protons in an oxide melt (see section 5.5 on Toop and Samis). However, Burnham showed that it may be used to calculate the solubilities of H_2O in natural melts of composition ranging from basalt to granitic pegmatite. To do this the compositions of the natural melts were expressed according to the stoichiometry of $NaAlSi_3O_8$. Thus, for example, diopside ($CaMgSi_2O_6$) becomes $Ca_{1.33}Mg_{1.33}Si_{2.67}O_8$ and SiO_2 becomes Si_4O_8. When

these recalculations are made, it is found that water
solubilities in a number of natural compositions agree
closely with that in $NaAlSi_3O_8$ when recalculated on the
basis of 8 oxygen atoms.

This implies that the components expressed as formulae
involving 8 oxygen atoms must mix close to _ideally_. Thus,
for example, the mixing of the components Si_4O_8, $NaAlSi_3O_8$,
$Ca_{1.33}Mg_{1.33}Si_{2.67}O_8$ etc. should be ideal. Activity-
composition relationships of this type may therefore be
used to calculate liquidus curves in the appropriate systems
using the methods described in sections (5.3) and (5.5).

Although this mixing model remains to be investigated
more fully, its apparent success in providing a simple
basis for considering the mixing behaviour of natural melts
may make it a most useful tool in igneous petrology.

6. Behaviour of trace components

6.1 INTRODUCTION

In the preceding chapters we have been primarily concerned
with the major components of rock-forming minerals and
melts. In natural systems however many components are
present only in trace amounts (approximately < 0.1 percent).
Their low concentrations (and activities) often prevent
the formation of phases of which they are major constituents
and they must therefore be accommodated as minor
components in mineral solid solutions, in melts or in other
fluid phases.

In minerals, trace components may be present as :

(i)   occluded zones trapped during rapid
       crystallization;
(ii)   interstitial defects in the host lattice;
(iii) solid solutions substituting for atoms of
       the host phases.

Although recent measurements appear to show that defect
formation may be important at very low trace levels, this
mechanism will be rapidly saturated in most geological

examples. In most cases, therefore, trace components in
minerals are present as solid solutions. It is important
however, to remember that 'minor' components may be major
constituents of some phases as in the cases of Zr in $ZrSiO_4$
or Cr in $FeCr_2O_4$.

The properties (e.g. size, charge, possible ligand field
stabilization) of trace impurities are frequently quite
different from those of the major constituents of the host
phase. Consequently the mixing behaviour of trace components
in minerals and melts is often highly non-ideal and trace
elements may exhibit strong 'preferences' for one phase
(liquid, solid, or fluid) relative to the others present in
natural systems. In such cases the trace element might,
for example, be concentrated in a growing mineral phase
(e.g. Ni^{2+} in olivine) or remain in the melt as an
'incompatible' element (e.g. K, Rb, light rare earth elements
etc.).

The partitioning of trace components between minerals
and melts may result in changes in trace concentrations of
several orders of magnitude during crystallization or melting
events. A quantitative understanding of the distribution
behaviour of trace elements may, therefore, allow their use
as highly sensitive monitors of igneous evolution (e.g.
Gast 1968).

6.2 DILUTE SOLUTIONS AND HENRY'S LAW

Although we have noted above that many trace components
interact strongly with other components in solution, their

low concentrations lead to simple activity-composition
relationships.

In an ideal solution there is no enthalpy of mixing
($H_{mix} = 0$; $S_{mix} = S_{config}$). In this case the activities of
components mixing on one site are given by Raoult's law :

$$a_i = X_i. \tag{6.1}$$

If the mixing components interact with one another
($H_{mix} \neq 0$), the activities will depart to a greater or
lesser extent from the ideal mixing curve (Fig. 6.1).
However, with increasing dilution, 'trace' components
become so dispersed that they are surrounded by a uniform
environment. Although such trace components may interact
strongly with the other components present, small changes

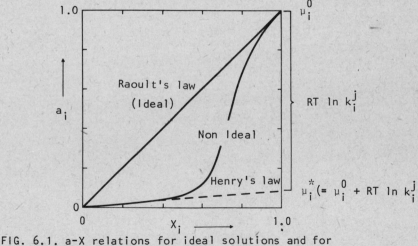

FIG. 6.1. a-X relations for ideal solutions and for
components obeying Henry's law.

in concentrations of a trace component do not significantly
affect the average environment because of the low concentrations involved. The activity coefficient therefore remains
constant and the activities of trace components thus become
directly _proportional_ to their concentrations (Fig. 6.1).

$$a_i^j = k_i^j \, x_i^j.$$ (6.2)

This is known as _Henry's law_, and the proportionality
constant k_i^j for the component i in phase j is the Henry's
law constant. The Henry's law constant is the activity
coefficient for species i in phase j at _high dilution_.

Some insight into the nature of the Henry's law constant
may be gained by considering a solution of i in equilibrium
with its vapour, where the latter is a perfect gas. At
equilibrium :

$$\mu_i^{vap} = \mu_i^{soln}$$ (6.3)

and

$$\mu_i^0 + RT \ln a_i^{vap} = \mu_i^0 + RT \ln a_i^{soln}$$ (6.4)

where μ_i^0 refers to the standard state of perfect gas at 1 bar.

Note from Fig. 6.1 that an extrapolation of the linear
(Henry's law) region can be constructed as shown. The
extension of the straight line intersects the ordinate for
pure (i.e. no longer trace) component i where $a_i \neq 1$.
Since the activity of pure component i is not 1, it is not

in its standard state (perfect gas). The deviation from the perfect gas standard state could be expressed as an activity coefficient term. Alternatively, the difference can be added to μ_i^0 to define a new standard state μ_i^* - the hypothetical Henry's law standard state - (compare the inclusion of $\int V \, dP$ in $\mu_{1 \text{ bar},T}^0$ to give $\mu_{P,T}^0$, Chapter 2). Using this standard state, the trace component obeys Raoult's law because all interactions have been included in μ_i^* (see Fig. 6.1) :

$$\mu_i = \mu_i^0 + RT \ln X_i^j \, k_i^j$$

$$= \mu_i^* + RT \ln X_i^j.$$

Thus, from eqns (6.3) and (6.4),

$$\underbrace{\mu_i^0 + RT \ln P_i}_{\text{perfect gas}} = \underbrace{\mu_i^* + RT \ln X_i}_{\text{solution}} \qquad (6.5)$$

and so

$$P_i/X_i = \exp (\mu_i^* - \mu_i^0)/RT. \qquad (6.6)$$

Since Henry's law for a gas in solution is

$$P_i = k_i X_i \qquad (6.7)$$

(cf. (6.2)) then

$$k_i = \exp (\mu_i^* - \mu_i^0)/RT. \qquad (6.8)$$

The Henry's law constant is thus related to the difference
in chemical potential between an ideal gas standard state
(μ_i^0) and the uniform-interaction extrapolated Henry's law
standard state (μ_i^*) (cf. Fig. 6.1). Since μ_i^* depends on the
strength of the interactions of the trace component with its
surroundings, the Henry's law constants for most components
are likely to change with the composition of the host phase
(section 3.11). These effects have usually been neglected
in the geological literature. However, recent experimental
work suggests that the compositional dependence of k_i may be
significant (Drake and Weill 1975, Sun et al. 1974) and that
for trace elements in different redox states, e.g. Eu^{2+}/Eu^{3+},
the compositional effects may be very important indeed
(Morris and Haskin 1974, Fraser 1975a).

6.3 DISTRIBUTION COEFFICIENTS

The simple a-X relationships in the Henry's law region
may be used to describe the distribution of trace elements
between different phases. For example, Häkli and Wright
(1967) have measured the distribution of Ni between olivines
and quenched liquids in samples obtained at known temperatures from the Makaopuhi lava lake in Hawaii, and Leeman
(1973) has investigated the system experimentally. At
equilibrium

$$\mu_{Ni_2SiO_4}^{melt} = \mu_{Ni_2SiO_4}^{ol} \qquad\qquad (6.9)$$

and

$$\mu^0_{Ni_2SiO_4,melt} + RT \ln a^{melt}_{Ni_2SiO_4} = \mu^0_{Ni_2SiO_4,ol} +$$
$$+ RT \ln a^{ol}_{Ni_2SiO_4}. \tag{6.10}$$

Applying Henry's law (6.2), this becomes

$$-\Delta G^0 = RT \ln \frac{k^{ol}_{Ni} \cdot C^{ol}_{Ni}}{k^{melt}_{Ni} \cdot C^{melt}_{Ni}} = RT \ln K \tag{6.11}$$

and

$$K. \frac{k^{melt}_{Ni}}{k^{ol}_{Ni}} = \frac{C^{ol}_{Ni}}{C^{melt}_{Ni}} = K_D \tag{6.12}$$

where C^{ol}_{Ni} and C^{melt}_{Ni} are concentrations in p.p.m. (formula weights cancel).

K_D is the Nernst distribution coefficient. Note that it includes the Henry's law constants k^{ol}_{Ni} and k^{melt}_{Ni}.

The Nernst distribution coefficient was originally formulated (Nernst 1891) to account for the distribution of trace elements between crystals and aqueous solutions. Since the major component of any aqueous solution is always H_2O, the effects of bulk compositional variation on observed distribution coefficients (i.e. on activity coefficients or Henry's law constants) are minimized. However, natural rock systems provide a very much more condensed environment with high site potentials in which the bulk composition may vary by comparatively large amounts. The exact value of the Henry's law constant k_i for a trace component i in phase j may therefore vary according to the composition of the phase. This may be seen by comparing the results of Sun et al. (1974) and Drake and Weill (1975) for the distribution coefficients of Sr between plagioclase and melts, obtained over a range of temperatures but for different bulk compositions. The results for the two systems may be compared at $1150^{\circ}C$ (Table 6.1).

TABLE 6.1

Comparison of distribution coefficients of Sr between plagioclase and melts at $1150^{\circ}C$ for different bulk compositions. Errors represent 99 percent confidence limits from regressions of ln K versus 1/T (from Drake and Weill 1975)

	Sun et al.	Drake and Weill
$K_D = C_{Sr}^{plag}/C_{Sr}^{melt}$	2.01 ± 0.09	3.06 ± 0.21
$K_C = (C_{Sr}/C_{Ca})^{plag}(C_{Sr}/C_{Ca})^{melt}$	1.69 ± 0.17	1.42 ± 0.16

The distribution of Yb between garnet and melt presents a more extreme case. Schnetzler and Philpotts (1970) report a value of $K_D = 40$ based on studies of natural iron-rich garnets in a dacitic melt. More recently, Shimizu and Kushiro (1975) have obtained a value of $K_D = 4.03$ from experimental runs at $1275^{\circ}C$ and 30 kbar in the system diopside-pyrope-H_2O. It is clear that in this case the effects of bulk composition, pressure and temperature may be very important.

An attempt to minimize the compositional dependence was made by Henderson and Kracek (1927), who included a major component in the distribution coefficient as a reference or 'carrier'. For example, the distribution coefficient for Sr between plagioclase and melt may be expressed using Ca as a carrier :

$$K_C = \frac{(C_{Sr}/C_{Ca})^{plag}}{(C_{Sr}/C_{Ca})^{melt}} .$$

(6.13

This approach is equivalent to considering an exchange reaction of the form :

$$CaAl_2Si_2O_8 + SrAl_2Si_2O_8 \rightleftarrows SrAl_2Si_2O_8 + CaAl_2Si_2O_8 .$$

plagioclase   melt   plagioclase   melt

(major)   (trace)

Values of K_C for both sets of data are also shown in Table 6.1 and can be seen to show closer agreement than values of K_D. However, even where major element analyses are reported together with trace element data, a distribution coefficient of this form cannot remove all bulk composition effects. Ideally, of course, the equilibrium constant in terms of activities should be used and this has been attempted by a number of authors (e.g. Vaslow and Boyd 1952, McIntire 1963). However, there is at present a major obstacle to a rigorous thermodynamic approach; this is the lack of information on the activity-composition relationships of multicomponent silicate melts and of solids (Chapters 3 and 5).

Although trace element distribution coefficients are still largely empirical, the authors feel that the influence of bulk composition on K_C and K_D would be greatly elucidated if geochemists were to determine both trace and all major component contents of all the phases they study. Without knowledge of the gross compositions of each phase, correction for bulk composition effects is impossible.

6.4 TRACE ELEMENT GEOTHERMOMETERS AND GEOBAROMETERS

Knowledge of the temperature and pressure dependence of trace element distribution coefficients may enable them to be used as geothermometers or geobarometers. The results of Häkli and Wright (1967) for Ni distribution are shown in Table 6.2 and were expressed as simple Nernst distribution coefficients K_D.

Since

$$\Delta G^0 = \Delta H^0 - T\Delta S^0 = - RT \ln K, \qquad (2.40) \text{ and } (2.41)$$

then

$$\ln K = \frac{-\Delta H^0}{RT} + \frac{\Delta S^0}{R}$$

and so the temperature dependence of an equilibrium constant is given by

$$\left(\frac{\partial \ln K}{\partial 1/T}\right)_P = \frac{-\Delta H^0}{R} . \qquad (6.14)$$

TABLE 6.2

Distribution of Ni between olivine and quenched glasses in tholeiites from Hawaii. Data from Häkli and Wright (1967)

Sample :		A	B	C	D	E	
Temperature (^{o}C) :		1050	1070	1075	1120	1160	
Olivine	Ni	840	935	955	1310	1555	p.p.m.
	Mg	n.d.	n.d.	32.17	36.60	39.24	wt. %
Glass	Ni	50	57	60	87	115	p.p.m.
	Mg	1.9	4.2	4.0	6.3	8.2	wt. %
	K_D	16.80	16.40	15.92	15.06	13.52	
	K_C	-	-	1.98	2.59	2.83	

A plot of ln K versus 1/T is therefore linear (if $\Delta C_p = 0$); the data from Table 6.2 are plotted as ln K_D (where K_D includes Henry's law constants) versus 1/T in Fig. 6.2.

From Fig. 6.2 it may be seen that the Ni distribution coefficient is strongly temperature dependent and might be used as a geothermometer. However, it should be remembered that, in this form, the exact values of K_D may vary with bulk composition. Taking the rather extreme case of the Ni_2SiO_4 - Mg_2SiO_4 phase diagram (Ringwood 1956) values of K_D in this system are all less than 1.0, Ni being enriched in the liquid phase. Using the Häkli and Wright data these

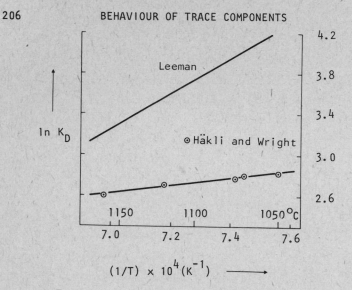

FIG. 6.2. Plot of $\ln K_D$ versus $1/T$ for Ni distribution coefficients Ni^{ol}/Ni^{liq}. Data from Häkli and Wright (1967) and Leeman (1973).

distribution coefficients imply temperatures of greater than infinity!

    More recently, Leeman (1973) has published some preliminary experimental measurements of the distribution of Ni between olivines and basaltic liquids. A line obtained from his regression,

$$\ln K_D = \frac{17\ 500}{T} - 9.0, \qquad\qquad (6.15)$$

is also shown in Fig. 6.2 From this equation it can be seen that Ni is enriched in the liquid phase above 1671°C.

Thus it seems that the true variation in K_D with temperature is much greater than that determined by Häkli and Wright from the natural rock samples.

The discrepancy between the two sets of results may be the result of kinetic disequilibrium during the growth of the natural olivines from the melt. This could result in <u>apparent</u> distribution coefficients having lower values than those measured at equilibrium (see section 6.7). Roeder and Emslie (1970) noted that the results of Häkli and Wright for Fe/Mg distribution between olivine and melt give calculated temperatures 45-65°C higher than the actual sample temperatures, and suggested that the melts were supersaturated with respect to olivine.

Irvine (1975) has considered both the temperature and compositional variation of K_D-values in a plot of ln K_D versus percent olivine in the norm (Fig. 6.3). With the

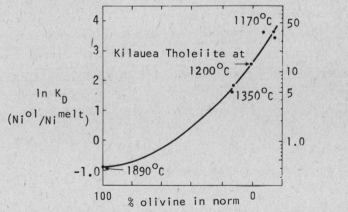

FIG. 6.3. Variation in Ni K_D-values with bulk composition. Data from Irvine (1975).

further development of thermodynamic models of silicate melts, it should become possible to account for the pressure-temperature-composition dependences of K_D using reactions of the form :

$$2NiO(melt) + Mg_2SiO_4(ol) = 2MgO(melt) + Ni_2SiO_4(ol) \quad (6.16)$$

$$K = \frac{a_{MgO}^{melt\,2} \cdot a_{Ni_2SiO_4}^{ol}}{a_{NiO}^{melt\,2} \cdot a_{Mg_2SiO_4}^{ol}} \cdot \quad (6.17)$$

As was seen in Chapter 5, the activities of oxide components in silicate melts change rapidly near the orthosilicate composition. The differences (probably as a result primarily of ligand-field effects) in a_{NiO} and a_{MgO} as a function of X_{SiO_2} can, in principle, be calculated using one of the polymer models described in Chapter 5. It will be of great interest to see if such calculations can account for the increased stabilization of NiO in very basic melts so that geothermometers of this sort may be given more general applicability. Irvine's (1975) plot of ln K_D versus percent olivine in the norm is a first empirical step towards this end.

6.5 RANGE OF APPLICABILITY OF HENRY'S LAW

Most trace-element distribution coefficients are based on the assumption that the trace component obeys Henry's law in the phases of interest.

The trace element concentrations up to which such an assumption is justified will vary from system to system (cf. the linear Henry's law region in Fig. 6.1). Experiments by Drake and Weill (1975) using different trace concentrations have shown that the distribution coefficients for Sr^{2+} and Ba^{2+} between plagioclase and melt are independent of compositions up to about 3 wt percent (30 000 p.p.m.). Grutzeck et al. (1973) have found a similar value for the limits of Henry's law behaviour for trivalent lanthanide ions between diopside and melt, while Leeman (1973) found that for NiO and CoO distribution between olivine and melt, Henry's law is obeyed up to about 20 and 15 wt percent in the melt respectively.

Thus it seems that in many cases Henry's law is obeyed to well above the concentration ranges observed in natural rocks.

6.6 TRACE COMPONENTS AS MONITORS OF IGNEOUS EVOLUTION

Because of the tendency of trace components either to be partitioned strongly into crystalline phases or into melts as 'incompatible' elements, trace element concentrations may vary by orders of magnitude in the crystallization of igneous rocks or during partial melting processes.

Following Schilling and Winchester (1967) and Gast (1968), a number of authors have examined the behaviour of trace components in some detail (e.g. Shaw 1970, Greenland 1970, Albarede and Bottinga 1972, Hertogen and Gijbels 1976, and a recent review is given by Arth (1976).

Crystallization models

The crystallization of minerals from melts may be considered in terms of three end-member processes :

1. Continuous re-equilibration with melt to give unzoned crystals (see section 6.7).

2. Surface equilibrium only :

   (a) Diffusion in crystal much slower than in melt or continuous removal of crystals.

   (b) Diffusion in melt much slower than in crystal.

In general, rates of diffusion in melts are very much higher than in crystals so that process 2(b) rarely, if ever, occurs in nature. In most cases either crystals are removed from the system by settling, or diffusion in the crystal is insufficiently rapid to maintain perfect equilibrium. In these cases process 2(a) will apply, and the limiting case is usually known as Rayleigh fractionation.

Rayleigh fractionation

Consider a finite magma reservoir containing a total of n moles of various components including y moles of a component Tr (e.g. Ni). Then the mole fraction of Tr in the system is $X_{Tr} = y/n$.

When a mineral containing Tr crystallizes, if each successive layer of crystals fails to maintain equilibrium with the remaining melt, either because of slow diffusion in the crystal or because of crystal settling then after a short time,

n becomes n-dn
y becomes y-dy.
$\qquad\qquad\qquad\qquad\qquad\qquad\qquad\qquad$ (6.18)

The compositions of the crystals and melt are now
given by :

$$X_{Tr}^{crystal} = \frac{dy}{dn}; \qquad X_{Tr}^{melt} = \frac{y-dy}{n-dn} \qquad\qquad (6.19)$$

If Tr is a trace component obeying Henry's law, then
from the distribution coefficient

$$X_{Tr}^{crystal} = \frac{dy}{dn} = K_D X_{Tr}^{melt} \qquad\qquad (6.20)$$

However, dy/dn can also be expressed directly in terms of
X_{Tr}^{melt}. From (6.19), $X_{Tr}^{melt} = (y-dy)/(n-dn)$. Neglecting dy
in comparison with y and dn in comparison with n,

$$X_{Tr}^{melt} \simeq \frac{y}{n} \text{ and so } y \simeq n.X_{Tr}^{melt}.$$

Differentiating both sides with respect to n, we obtain

$$\frac{dy}{dn} = n.\frac{dX_{Tr}^{melt}}{dn} + X_{Tr}^{melt}.$$

This expression for dy/dn may now be substituted into (6.20) :

$$K_D \, X_{Tr}^{melt} = n . \frac{dX_{Tr}^{melt}}{dn} + X_{Tr}^{melt}, \qquad (6.21)$$

and rearrangement of (6.21) yields the expression

$$\frac{1}{X_{Tr}^{melt}(K_D - 1)} . \, dX_{Tr}^{melt} = \frac{1}{n} . \, dn. \qquad (6.22)$$

The total changes in concentration during crystallization can now be obtained by integration of (6.22) between X_{Tr}^0 and X_{Tr} and between the initial and final amounts of melt n^0 and n.

$$\ln \frac{n}{n^0} = \frac{1}{K_D - 1} \ln \left(\frac{X_{Tr}^{melt}}{X_{Tr,melt}^0} \right)$$

and

$$\frac{X_{Tr}^{melt}}{X_{Tr,melt}^0} = \left(\frac{n}{n^0} \right)^{K_D - 1} . \qquad (6.23)$$

(n/n^0) is the fraction of original melt remaining, F. Thus we have

$$X_{Tr}^{melt} = X_{Tr,melt}^0 (F)^{K_D - 1} \qquad (6.24)$$

Equation (6.24) is the Rayleigh fractionation law (Rayleigh 1896).

According to this, as crystallization proceeds, the mole fraction of Tr in the melt at any time will increase or decrease according to $(F)^{K_D - 1}$. Plots of X_{Tr} versus $(F)^{K_D - 1}$ are shown in Fig. 6.4 for different values of K_D. This figure shows that it is _impossible_ by a simple fractionation process to produce greater enrichment in residual liquids than indicated by the curve for K = 0 (the most favourable case).

The enrichment trends for some incompatible trace elements in the Skaergaard intrusion (Wager and Mitchell 1951) are compared with theoretical curves for perfect fractional crystallization behaviour in Fig. 6.4. The theoretical and observed trends show marked similarity.

Equilibrium crystallization

If no crystal settling occurs and cooling is sufficiently slow and diffusion sufficiently rapid that the interiors of the crystals maintain equilibrium with the melts during crystallization, then end-member process (1) is obtained.

Consider the same magma reservoir as before containing a total of n^0 moles of various components of magma of which

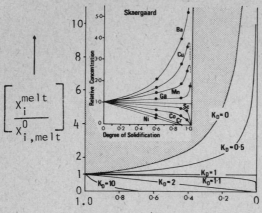

$$\left[\dfrac{x_i^{melt}}{x_{i,melt}^0}\right]$$

F (Fraction of melt remaining)

FIG. 6.4. Changes in trace element concentration in residual liquids during Rayleigh fractionation for different values of K_D.

y_{Tr}^0 are of a trace component Tr.

If equilibrium between crystals and melt is achieved at all stages of crystallization (and if K_D does not change) then

$$\dfrac{x_{Tr}^{solid}}{x_{Tr}^{melt}} = K_D \text{ in all cases.} \qquad (6.25)$$

Let a fraction $(1-F)$ of the magma crystallize so that the fraction of melt remaining is

$$F = \dfrac{n_{melt}}{n_{melt}^0} . \qquad (6.26)$$

The ratio of solid to melt remaining is therefore given by

$$\frac{\text{solid}}{\text{melt remaining}} = \frac{n^0_{melt} - n_{melt}}{n_{melt}} = (\frac{1}{F} - 1). \qquad (6.27)$$

We can use (6.27) to substitute for X_{Tr} in (6.25). Thus

$$\frac{X^{solid}_{Tr}}{X^{melt}_{Tr}} = \frac{n^{solid}_{Tr}/\text{total solid}}{n^{melt}_{Tr}/\text{melt remaining}} = \frac{y^{solid}_{Tr}/(n^0_{melt} - n_{melt})}{y^{melt}_{Tr}/n_{melt}} = K_D \qquad (6.28)$$

and since

$$y^{solid}_{Tr} = y^0_{Tr,melt} - y_{Tr,melt}. \qquad (6.29)$$

we obtain

$$K_D = \frac{y^0_{Tr,melt} - y_{Tr,melt}}{y_{Tr,melt}} \cdot \frac{n_{melt}}{n^0_{melt} - n_{melt}}$$

$$= \left(\frac{y^0_{Tr,melt}}{y_{Tr,melt}} - 1\right) \cdot \frac{n_{melt}}{n^0_{melt} - n_{melt}}.$$

Rearranging, and dividing by n^0_{melt}/n_{melt}, we obtain

$$\frac{y^0_{Tr,melt}/n^0_{melt}}{y_{Tr,melt}/n_{melt}} = \left[K_D \left(\frac{n^0_{melt} - n_{melt}}{n_{melt}}\right) + 1\right] \frac{n}{n^0} , \qquad (6.30)$$

and hence using (6.26)

$$\frac{X_{Tr,melt}}{X^0_{Tr,melt}} = \frac{1}{K_D \ (1-F) + F} \ . \tag{6.31}$$

This is the expression for changes in trace element concentrations during equilibrium crystallization of a magma. Plots of $X_{Tr,melt}/X^0_{Tr,melt}$ against the extent of crystallization (1-F) are shown in Fig. 6.5 for various values of K_D.

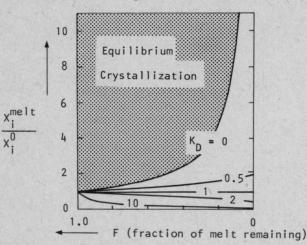

FIG. 6.5. Changes in concentrations of trace components in a melt during equilibrium crystallization for different values of K_D.

As an example of the use of these equations, it is of interest to consider how trace element concentrations may be applied to test possible petrogenetic relationships, for example between oceanic tholeiite basalts and oceanic alkali

basalts. Oceanic basalts are likely to have suffered a
minimum of crustal contamination and it is proposed that
the alkali basalts can be derived from a tholeiitic parental
magma by fractional crystallization. Average values of the
K and Rb contents of the two groups of rocks are :

	K	Rb
Oceanic tholeiites :	0.08 percent	1 p.p.m.
Alkali basalts :	0.9 percent	18 p.p.m.

Taking bulk distribution coefficients (cf. eqn (6.32))
for the whole rock to be 0.02 for both K and Rb, the extent
of crystal fractionation required to produce the observed
enrichments of these elements in alkali basalts can be
determined. In the most favourable case, the limiting
condition of Rayleigh fractionation will apply.

$$\frac{X^{melt}}{X^{melt}_0} = (F)^{K_D - 1} \tag{6.24}$$

The enrichment factors (X/X_0) for K and Rb between the
alkali basalts and tholeiites are 11.25 and 18.0 respectively.
It can be seen from eqn (6.24) that to produce such degrees of
enrichment, the trace elements must be concentrated into
fractions of melt remaining (F) of 0.0846 in the case of K
and 0.0524 for Rb. These values are equivalent to 91.5 and
94.8 percent crystallization.

Consideration of these trace element data therefore
provides a quantitative argument against a simple fractional
crystallization hypothesis. It is very difficult to imagine

any combination of phases which could be removed from a
tholeiitic melt so as to achieve 90 to 95 percent
crystallization without substantially altering the major
element chemistry of the residual liquid. Any such
hypothesis must also be able to explain the change in the
K/Rb ratio.

As might be expected, the less efficient case of
equilibrium crystallization requires concentration into still
smaller volumes of melt and use of eqn (6.31) yields values
of 93.0 and 96.4 percent crystallization to produce the
observed degrees of K and Rb enrichment.

6.7 PARTIAL MELTING MODELS

The general principles of trace element behaviour
discussed above may also be applied to constrain the extent
of trace-element enrichment or depletion during partial
melting events. Following the work of Gast (1968), three
types of melting process have been considered in detail
(Shaw 1970, Greenland 1970, Albarede and Bottinga 1972,
Hertogen and Gijbels 1976). These are :

1. Continuous re-equilibration of melt with
    residual solid phases throughout partial
    melting until removal of melt (batch melting).

2. Continuous removal of infinitesmally small
    increments of partial melt from residual solids -
    fractional melting.

3. Continuous removal of partial melt from residual
    solids. This process differs from 2. above in that
    the melt collects in a common well-mixed magma
    chamber so as to give a final averaged composition.

One of the underlying assumptions made in the geological
literature when discussing these different melting models is
that the distribution coefficients remain <u>constant</u> throughout
each melting event. Although there is accumulating evidence
that this is not so, as has been discussed above, the
consequences of these models may be reviewed at present
using the usual assumption of constant K_D values. Consider a
rock consisting of the phases α, β, γ, etc. The bulk
distribution coefficient of such a rock may be expressed as
the weighted mean of the individual mineral distribution
coefficients for component i. Thus following Gast (1968) and
Shaw (1970), we have for each component

$$D_0^{rock} = X_0^\alpha K_D^{\alpha/1} + X_0^\beta K_D^{\beta/1} + X_0^\gamma K_D^{\gamma/1} + \ldots \qquad (6.32)$$

where D_0^{rock} is the distribution coefficient for the starting
assemblage, $K_D^{\alpha/1}$, $K_D^{\beta/1}$ etc. are the crystal-liquid
distribution coefficients for the component between each
mineral and initial melts, and X_0^α, X_0^β etc. are the initial
weight fractions of each phase.

If the distribution coefficients $K_D^{\alpha/1}$ etc. do not change
with changing melt composition and temperature during a
partial melting event, then letting F be the fraction of melt
produced, the different melting models yield :

1. Batch melting

$$\frac{C_i^{melt}}{C_i^0 \; (solid)} = \frac{1}{D_0 + F \; (1-P)} \qquad\qquad (6.33)$$

where C_i^0 was the initial concentration of component i in the rock, and

$$P = (p^\alpha \; K_D^{\alpha/1} + p^\beta \; K_D^{\beta/1} + p^\gamma \; K_D^{\gamma/1} + \ldots)$$

in which p^α, p^β etc. are the fractions of liquid contributed by each phase.

If the phases melt according to their relative modal proportions in the rock (i.e. when X^α, X^β etc. do not change during melting), then (6.33) simplifies to the (unusual) case of modal melting :

$$\frac{C_i^{melt}}{C_i^0 \; (solid)} = \frac{1}{D_0 + F \; (1-D_0)} \; . \qquad\qquad (6.34)$$

The changes in concentration during a batch melting process for which all phases melt according to their proportions in the rock (modal melting) are shown in Fig. 6.6. Note again that the maximum enrichment for any given degree of partial melting cannot exceed that described by the curve for $D_0 = 0$.

In general the minerals in a rock will not melt according to their modal proportions nor will the contribution of each phase to the amount of melt remain constant

$(p^{\alpha}, p^{\beta}$ etc.). Some of the more complex equations required to express this behaviour have been recently developed by Hertogen and Gijbels (1976).

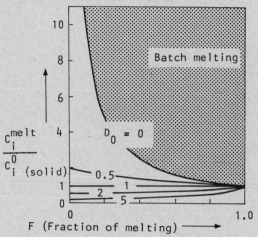

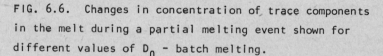

FIG. 6.6. Changes in concentration of trace components in the melt during a partial melting event shown for different values of D_0 - batch melting.

2. Continuous removal of melt - fractional melting

In this case the Rayleigh fractionation law leads to

$$\frac{c_i^{melt}}{c_i^0 \text{ (solids)}} = \frac{1}{D_0} \left(1 - \frac{PF}{D_0}\right)^{(1/P - 1)}$$

(6.35)

and for modal melting this becomes

$$\frac{c_i^{melt}}{c_i^0 \text{ (solids)}} = \frac{1}{D_0} (1 - F)^{(\frac{1}{D_0} - 1)} .$$

(6.36)

It is most unlikely that the removal of infinitesimally small amounts of melt can be accomplished in nature, so that these equations represent limiting conditions for melting processes.

3. Continuous removal and collection of melt

If the separating melts aggregate together in a well-mixed reservoir, then the concentration of trace element in the melt is given by

$$\frac{\bar{c}_i^{melt}}{c_i^0 \text{ (solids)}} = \frac{1}{F} \left[(1 - \frac{PF}{D_0})^{1/P} \right]$$

(6.37)

and for modal melting :

$$\frac{\bar{c}_i^{melt}}{c_i^0 \text{ (solids)}} = \frac{1}{F} \left[1 - (1 - F)^{1/D_0} \right]$$

(6.38)

$\bar{c}_i^{melt}$ is the averaged concentration of the trace element in the mixed melt.

A comparison of the results of these three models is shown in Fig. 6.7. In geologically reasonable cases, it is likely that an appreciable batch of magma must accumulate before its separation from residual phases. However, the

likely extent of equilibration between such a batch and
the residual rocks, and the minimum size needed for
separation remain unknown.

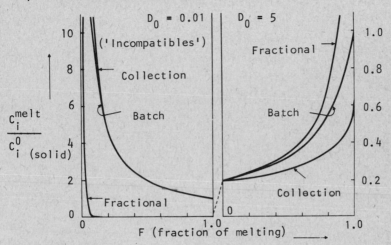

FIG. 6.7. A comparison of fractional melting, batch
melting and removal plus collection (see text) models for
two possible values of bulk distribution coefficient.

It was stressed above that these calculations assume
that individual distribution coefficients do not vary with
changing composition, temperature or pressure. This is
most unlikely to be true, and Hertogen and Gijbels (1976)
have developed some of the more complex equations needed to
allow for variations in K_D. Since precise experimental
data describing the exact nature of the variations of K_D
with T, P, and X are only beginning to become available,
these more complex cases will not be considered here.

6.8   APPARENT DISTRIBUTION COEFFICIENTS

Many of the distribution coefficients quoted in the literature have been obtained from natural rocks by bulk analysis of phenocryst and matrix materials. In the case of equilibrium crystallization discussed above, this procedure should yield the correct value. However in cases where minerals are zoned, due either to slow diffusion in the crystal (Rayleigh fractionation) or in the melt, measurement of the bulk distribution coefficient yields an incorrect value (Albarede and Bottinga 1972).

In cases where $K_D > 1$ and the Rayleigh fractionation law is obeyed, 'apparent' distribution coefficients will be greater than equilibrium (Nernst) coefficients. If crystal zonation is the result of slow diffusion in the melt towards the growing crystal, 'apparent' distribution coefficients will have anomalously low values.

Great caution should therefore be exercised in the use of trace element distribution coefficients obtained from bulk analysis of co-existing phases. Nevertheless these may reflect the behaviour of real geological systems more accurately than the true equilibrium values.

SUMMARY - CHAPTER 6

1. In dilute solution, trace components may obey Henry's law ($a_i^j = k_i^j X_i$).

2. The Henry's law constant k_i^j for a component i in phase j is a function of T, P, and the composition of phase j.

3. Trace element concentrations may vary enormously during
   crystallization or partial melting processes. The exact
   amount will depend on the magnitude of the distribution
   coefficients, the type of crystallization or partial
   melting, and any variations in distribution coefficients
   during the process.

4. This chapter has been concerned primarily with the
   principles governing the behaviour of trace components
   which do not form phases of their own. The application
   of these principles to rocks is a field in which rapid
   advances are currently being made. However, the general
   conclusions which emerge from these studies are that it
   is difficult to change the ratios of 'incompatible' trace
   components by simple fractionation models and that in
   many cases the extent of crystallization required to give
   the required degree of enrichment is too high to be
   accomplished without fundamentally altering the composition
   of the evolving magma. The Interested reader is
   referred to Gast (1968), Kay and Gast (1973), O'Nions and
   Pankhurst (1973) and Sun and Hanson (1975) amongst others
   for more detailed discussions of the field.

PROBLEMS - CHAPTER 6

1.   Shimizu and Kushiro (1975) have reported an experimental value of K_D = 4.03 for Yb between garnet and melt
   in the system diopside-pyrope-H_2O at $1275^{o}C$, 30 kbar.
   O'Hara (1968) has proposed that the higher K_2O content of

alkali basalts compared to tholeiite may be explained
by a small degree of partial melting followed by
extensive eclogite fractionation. If weighted mean
crystal-melt distribution coefficients, D_0, for garnet-
clinopyroxene assemblages are taken as 2.0 for Yb and
0.1 for K^+ :

(a) How much crystallization of eclogite would be
necessary to increase the K^+ content of residual liquids
by a factor of 2, assuming Rayleigh fractionation?

(b) How much would the Yb content of the melts have
decreased during this process?

(c) Repeat the calculations assuming equilibrium
crystallization.

SOLUTIONS TO PROBLEMS - CHAPTER 6

1(a) For Rayleigh fractionation, the concentration of trace
elements in residual melts is given by (6.24) :

$$X_{Tr} = X_{Tr}^0 \ (F)^{K_D - 1} \tag{1}$$

where F is the fraction of melt remaining. For any given
increase in concentration (X_{Tr}/X_{Tr}^0), the extent of crystal
fractionation necessary (1-F) is obtained from :

$$\ln X/X^0 = (K_D - 1) \ \ln . F. \tag{2}$$

In the present case, for an increase in potassium content by a factor of two, we have

$$\ln 2 = (0.1 - 1) \ln F$$

and so $F = 0.46$.

Thus to increase the content of K_2O in a residual liquid, 54 percent crystallization of eclogite would be required.

1(b) The change in Yb content can now be obtained from (1) :

$$\frac{X_{Yb}}{X_{Yb}^0} = (0.46)^{1.0} = 0.46.$$

The importance of using the correct value for K_D can be illustrated by comparing this result with that obtained using the value of 40 for K_D obtained by Schnetzler and Philpotts (1970) for garnets in a dacitic liquid. Using a weighted mean distribution coefficient of 20 for Yb, we have

$$\frac{X_{Yb}}{X_{Yb}^0} = (0.46)^{(20-1)} = 3.9 \times 10^{-7}.$$

In this case, extreme depletion of Yb would be expected for 54 percent fractional crystallization.

1(c) For equilibrium crystallization, we have from (6.31)

$$\frac{X_{Tr,melt}}{X^0_{Tr,melt}} = \frac{1}{K_D(1-F)+F} \quad .$$
(3)

Solving for F, we obtain

$$\frac{(X^0_{Tr,melt}/X_{Tr,melt}) - K_D}{1 - K_D} = F$$
(4)

and so for a two-fold increase in potassium content we have :

$$F = \frac{0.5 - 0.1}{0.9} = 0.44.$$

This is equivalent to $(1-F) = 56$ percent crystallization. The corresponding depletion in Yb is given by

$$\frac{X_{Yb,melt}}{X^0_{Yb,melt}} = \frac{1}{2(0.56) + 0.44} = 0.64.$$

Thus, in the case of equilibrium crystallization Yb is less depleted (to 0.64 of its initial value) than in the other limiting case of Rayleigh fractionation.

7. Estimation of thermodynamic data

7.1 INTRODUCTION

The usefulness of the thermodynamic approach to the study
of reactions in natural systems as opposed to direct
experimentation is twofold.   Firstly, provided thermodynamic data are available, it is possible to calculate the
positions of reactions which cannot be studied experimentally
because of the time necessary to approach equilibrium. This
applies particularly to reactions involving solids only and
reactions which take place at very low temperatures. In
addition, provided activity-composition relationships for
mixed solid, fluid, and melt phases are available, it is
possible to calculate the effects of variable composition of
any phase on the reaction of interest.   To attempt to do
this by performing experiments on every conceivable bulk
composition and analysing product phases over all pressure-
temperature space is clearly not feasible. The validity of
the thermodynamic approach hinges, of course, on the
availability of enthalpy, entropy, and volume data for pure
phases and a knowledge of the mixing properties of multicomponent phases.   The available data on mixing in solid,

fluid, and melt phases has been discussed in Chapters 3
and 5. This chapter is devoted to thermodynamic data for
pure phases.

Robie and Waldbaum (1968) give a comprehensive
tabulation of enthalpy, entropy, and volume data for most
minerals of geological interest. These data were obtained
from calorimetric studies and from phase equilibrium experiments in the manner described in section 7.4. The Robie and
Waldbaum data are, in many cases, adequate for one to perform
reasonably accurate calculations of equilibria in simple and
complex systems. In some cases, however, errors arise
because not all of the enthalpy data are internally
consistent. The internal inconsistencies have arisen
because of discrepancies between the accepted enthalpy
values of the 'standard' compounds used by the experimentalists performing the calorimetric studies. In particular,
discrepancies in data for aluminium-bearing compounds, now
to some extent resolved (e.g. Zen, 1972, Thompson, 1974),
have arisen due to errors in the measured enthalpies of
formation of corundum (Al_2O_3) and gibbsite $(Al(OH)_3)$.

7.2   CALORIMETRIC STUDIES

The principle behind the derivation of enthalpy data
for silicates is extremely simple and based on the additivity
of enthalpies discussed in Chapter 1. It is not possible,
nor is it necessary to obtain absolute values of the heat
contents of phases. It is sufficient to obtain values
relative to elements or oxides and these may be derived

calorimetrically by accurate measurement of the heat
evolved when the silicates and their constituent oxides
are each dissolved in a suitable solvent. The most common
solvent used for these studies is hydrofluoric acid, HF. In
recent years lead borate melts have also been used.

The enthalpy of the reaction

$$2MgO \quad + \quad SiO_2 \; \underset{\leftarrow}{\rightarrow} \; Mg_2SiO_4 \qquad \Delta H_1. \tag{7.1}$$
periclase   quartz   forsterite

is quite large (- 15.1 kcal (- 63.2 kJ) at 1 bar, 298.15 K)
but cannot be measured directly because the reaction, in
common with all solid-solid reactions, takes place very
slowly. However, the enthalpy of reaction can be determined
by measuring the heats of solution of MgO, SiO_2, and
Mg_2SiO_4 in HF at low temperatures (about $60^{\circ}C$).

$$2MgO \quad + \quad SiO_2 \; + \quad HF \; \underset{\leftarrow}{\rightarrow} \; (2MgO,SiO_2) \; \Delta H_2. \tag{7.2}$$
periclase   quartz   solution   solution

$$Mg_2SiO_4 \; + \quad HF \; \underset{\leftarrow}{\rightarrow} \; (2MgO,SiO_2) \; \Delta H_3. \tag{7.3}$$
forsterite   solution   solution

Making the assumption that the product solution is identical
in both cases, it can be seen that reaction (7.1) is
equivalent to reaction (7.2) minus reaction (7.3), so that
ΔH_1 is simply given by

$$\Delta H_1 = \Delta H_2 - \Delta H_3. \tag{7.4}$$

A considerable body of data on the enthalpies of formation of compound silicates from their constituent oxides has been built up by analogous calorimetric methods. It should be noted that, in calorimetric studies, enthalpies of formation are generally quoted relative to the oxides as in reaction (7.1), whereas the tables of Robie and Waldbaum (1968) and some more recent tabulations (e.g. Zen 1972, Thompson 1974) give enthalpies of formation relative to the constituent elements of the compound.

The enthalpy of formation of forsterite from Mg, Si and O_2 at any temperature may be obtained by combining ΔH_1 with the enthalpies of the two reactions :

$$Mg + \tfrac{1}{2}O_2 \; \underset{\leftarrow}{\rightarrow} \; MgO \qquad \qquad \Delta H_5 \qquad \qquad \qquad (7.5)$$
$$\text{periclase}$$

$$Si + O_2 \; \underset{\leftarrow}{\rightarrow} \; SiO_2 \qquad \qquad \Delta H_6 \qquad \qquad \qquad (7.6)$$
$$\text{quartz}$$

where Mg, Si, and O are in their stable forms at the temperature of interest.

Consider the net reaction :

$$2Mg + Si + 2O_2 \; \underset{\leftarrow}{\rightarrow} \; Mg_2SiO_4 \qquad \Delta H_7. \qquad \qquad (7.7)$$

The enthalpy of formation of forsterite from the elements, ΔH_7, denoted H_f hereafter, is given by

$$(H_{forst} - 2H_{Mg} - H_{Si} - 2H_{O_2}) = \Delta H_7 = H_f = \Delta H_1 +$$

$$+ 2\Delta H_5 + \Delta H_6. \qquad (7.8)$$

At 298.15 K, $H_{f, 298.15}$ is - 520 370 $\pm$ (520) cal (- 2177.2 $\pm$ (2.2) kJ) per mole of forsterite formed. This value is made up of the following values :

$$\Delta H_1 = - 15\ 120\ cal; \quad 2 \times \Delta H_5 = 2 \times - 143\ 800\ cal;$$

$$\Delta H_6 = - 217\ 650\ cal.$$

The enthalpies of the reactions between the elements, ΔH_5, ΔH_6, and so on, have in most cases been determined calorimetrically by direct reaction at high temperatures. Tabulated heat of formation data may be used directly to calculate the enthalpies of reactions of interest. Consider the reaction :

$$\underset{\text{forsterite}}{Mg_2SiO_4} + \underset{\text{quartz}}{SiO_2} \rightleftarrows \underset{\text{clinoenstatite}}{Mg_2Si_2O_6} \qquad \Delta H_9. \qquad (7.9)$$

$$(H_f)_{forst} = (H_{forst} - 2H_{Mg} - H_{Si} - 2H_{O_2}) \qquad (7.10a)$$

$$(H_f)_{qz} = (H_{qz} - H_{Si} - H_{O_2}) \qquad (7.10b)$$

$$(H_f)_{clinoenst} = (H_{clinoenst} - 2H_{Mg} - 2H_{Si} - 3H_{O_2}) \qquad (7.10c)$$

Subtracting (7.10a) and (7.10b) from (7.10c) gives

$$(H_f)_{clinoenst} - (H_f)_{forst} - (H_f)_{qz} = (H_{clinoest} - H_{qz} - H_{forst})$$

$$= \Delta H_9.$$

From eqn (7.10) it may be seen that the enthalpy of the reaction is the sum of the enthalpies of formation of product phases, minus the sum of the enthalpies of formation of the reactants.

The difficulty with using tabulations of enthalpy data is that some of the data are internally inconsistent, as has already been discussed. Before using any such data it is suggested, therefore, that the reader check (as far as possible) the consistency of the values to be used with whatever phase equilibrium studies are available. Alternatively, direct extraction of ΔH^0 and ΔS^0 from phase equilibrium studies may be the most accurate approach (see section 7.4).

7.3 PHASE EQUILIBRIUM EXPERIMENTS AND THERMODYNAMIC DATA - INTRODUCTION

Although tabulated heat of formation data may be used to calculate the enthalpies of reactions it is not, in most cases, necessary to know the heats of formation of the individual phases involved in the reactions of interest.

Let us suppose, for example, that enthalpy and entropy data are required for the reaction

$$3CaAl_2Si_2O_8 \rightleftarrows Ca_3Al_2Si_3O_{12} + 2Al_2SiO_5 + SiO_2. \qquad (7.11)$$

anorthite   garnet   kyanite   quartz

This reaction, with all phases pure, has been studied
experimentally by Hays (1966) and Hariya and Kennedy (1968).
At each point on the reaction boundary between low pressure
(anorthite) and high pressure (garnet, kyanite, quartz)
phases, the condition of equilibrium is

$\Delta G = 0.$

Taking the standard states to be the pure phases at the
pressure and temperature of interest we have

$$\Delta G = \Delta G^0 = \mu^0_{Ca_3Al_2Si_3O_{12}} + 2\mu^0_{Al_2SiO_5} + \mu^0_{SiO_2} - \quad (7.12)$$

$$- 3\mu^0_{CaAl_2Si_2O_8} = (G^0_{gt} + 2G^0_{ky} + G^0_{qz} - 3G^0_{an}) = 0.$$

Assuming ΔV^0 independent of pressure and temperature, we
have

$$\Delta G^0 = 0 = \Delta H^0_{1\ bar,T} - T\Delta S^0_T + (P-1)\ \Delta V^0. \qquad (7.13)$$

Making the general assumption for solid-solid reactions
that $\Delta H^0_{1\ bar,T}$ and ΔS^0_T are independent of temperature, and
given volumes of all phases (e.g. from crystallographic
data), it is strictly only necessary to know two values of
P and T at which $\Delta G^0 = 0$ in order to obtain $\Delta H^0_{1\ bar,T}$ and
ΔS^0_T. If experimental pressures and temperatures are known
for two equilibrium points, then from (7.13) we have two
equations with two unknowns ($\Delta H^0_{1\ bar,T}$ and ΔS^0_T) which can

be solved simultaneously. In cases where the reaction of
interest has been studied experimentally, therefore,
enthalpy and entropy values may be obtained directly from the
experiments, and heat of formation data are not required. The
results may then be applied to multicomponent phases using
appropriate activity-composition relationships.

  If the reaction of interest has not been studied experimentally, then it may be possible to obtain ΔH and ΔS from
experiments on related reactions using the principle of
additivity of enthalpy and entropy. For example, let us
suppose that a natural assemblage of garnet, plagioclase,
quartz, and andalusite (not kyanite) is under study. A
reaction analogous to (7.11) may be written for this
assemblage as follows :

$$3CaAl_2Si_2O_8 \rightleftarrows Ca_3Al_2Si_3O_{12} + 2Al_2SiO_5 + SiO_2. \qquad (7.14)$$
$$\text{anorthite} \qquad \text{garnet} \qquad \text{andalusite} \quad \text{quartz}$$

This reaction cannot be studied experimentally in the simple
'pure' system because pure anorthite does not break down at
pressures within the andalusite stability field. In
complex natural systems, however, the assemblage
plagioclase$_{ss}$, garnet$_{ss}$, andalusite, quartz, may occur.
To obtain enthalpy and entropy data for reaction (7.14), it
is necessary only to obtain the values for reaction (7.11)
(ΔH_{11} and ΔS_{11}) and to derive enthalpy and entropy data for
the reaction

$$Al_2SiO_5 \;\rightleftarrows\; Al_2SiO_5 \qquad\qquad \Delta H_{15}, \; \Delta S_{15} \qquad\qquad (7.15)$$
kyanite   andalusite

The enthalpy and entropy changes of reaction (7.14) may now be obtained by addition :

$$\Delta H_{14} = \Delta H_{11} + 2\Delta H_{15} \qquad\qquad\qquad (7.16)$$

$$\Delta S_{14} = \Delta S_{11} + 2\Delta S_{15}. \qquad\qquad\qquad (7.17)$$

7.4 EXAMPLE OF THE EXTRACTION OF THERMODYNAMIC DATA FROM PHASE EQUILIBRIUM EXPERIMENTS

Consider the reaction

$$Ca_3Al_2Si_3O_{12} + SiO_2 \;\rightleftarrows\; CaAl_2Si_2O_8 + 2CaSiO_3 \qquad (7.18)$$
grossular   quartz   anorthite   wollastonite

$$\Delta V^0_{298,1\ bar} = + 32.66\ cm^3 = 0.7806\ cal\ bar^{-1},$$

which has been studied experimentally by Newton (1966) and Boettcher (1970).

Shown in Fig. 7.1 are most of the experimental data which may be used to define the stability fields of the two assemblages in the simple system $CaO-Al_2O_3-SiO_2$.

As is to be expected, in view of the positive volume change of reaction, grossular + quartz is the high-pressure assemblage. Anorthite and wollastonite are stable together at high temperatures and low pressures.

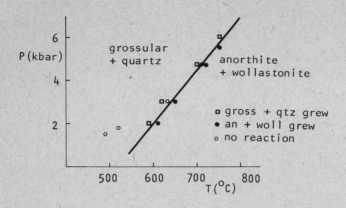

FIG. 7.1. The stability fields of grossular + quartz and
anorthite + wollastonite. Experimental data from Newton
(1966) and Boettcher (1970).

At equilibrium, taking standard states of all components
to be the pure phase at the pressure and temperature of
interest, we have :

$$\Delta G^0_{P,T} = 0 = \Delta H^0_{1\ bar,T} - T\Delta S^0_T + \int_1^P \Delta V^0\ dP. \qquad (7.19)$$

If we make the assumption that ΔC_p is 0 and that ΔV^0 is
constant, two points on the equilibrium boundary would be
enough to obtain $\Delta H^0_{1\ bar}$ and ΔS^0. However, as can be seen
from Fig. 7.1, phase 'equilibrium' experiments are
performed in such a way that equilibrium between products
and reactants is rarely obtained. The experimentalist

mixes product and reactant assemblages together and then
holds the mixture at a known pressure and temperature for an
appropriate length of time (usually days or weeks). Some
experiments on the assemblage of (7.18) result in the growth
of anorthite and wollastonite, others in the growth of
grossular and quartz. It is rare for no reaction to be
observed. Rather than determining the actual position of the
equilibrium boundary, therefore, it is 'bracketed' by
experiments in which growth of products takes place and others
in which reactants grow. It is from these 'bracketing'
points that thermodynamic data must be derived.

An important point which must be made here is the
significance of experiments in which no reaction is observed.
If the pressure and temperature of the experiment is close to,
or on, the equilibrium boundary, then no reaction should take
place. It is tempting, therefore, to assume that 'no
reaction' corresponds to equilibrium between products and
reactants and to use the results to obtain enthalpy and
entropy data. Fig. 7.1 illustrates the inadequacy of this
assumption for, although two experiments which yielded no
reaction lie close to the equilibrium boundary, two others
are nearly 100°C away from it. The reason for the errors in
the latter points is that at low temperatures reaction rates
are slow and little reaction may be observed in long experiments even under conditions far removed from equilibrium.
'No reaction' is not, therefore, an adequate criterion of
equilibrium. Only those experimental runs in which reaction
definitely occurred may be used to extract data.

An experiment in which anorthite and wollastonite were produced from grossular and quartz has (assuming $\Delta C_p = 0$ and ΔV constant) the conditions (from Fig. 1.1) :

$$\Delta G^0_{P,T} = \Delta H^0_{1\ bar} - T\Delta S^0 + (P-1)\ \Delta V^0 < 0 \text{ (products stable)}. \quad (7.20)$$

A run in which grossular and quartz formed has the condition

$$\Delta H^0_{1\ bar} - T\Delta S^0 + (P-1)\ \Delta V^0 > 0 \text{ (reactants stable)} \quad (7.21)$$

Thus, for each 'bracketing' run an inequality of this type can be set up for the pressure and temperature at which the experiment was performed (see Gordon 1973). The experiments shown in Fig. 7.1 in which reaction occurred are given in Table 7.1, together with the expression

$$\Delta H^0_{1\ bar} - T\Delta S^0 \gtrless - (P-1)\ \Delta V^0. \quad (7.22)$$

The greater-than or less-than signs were obtained from the directions in which reaction was observed to take place.

For each run given in Table 7.1 it is possible to plot a line on a graph of $\Delta H^0_{1\ bar}$ against ΔS^0 which corresponds to the condition

$$\Delta H^0_{1\ bar} - T\Delta S^0 = - P\Delta V^0, \quad (7.23)$$

and, depending on whether products or reactants formed, the experimental result constrains the internally consistent values of $\Delta H^0_{1\ bar}$ and ΔS^0 to lie on one side or the other

TABLE 7.1

Experimental results for the reaction:

gross + qz $\rightleftarrows$ an + woll

	P(kbar)	T(K)	Result			$- P\Delta V^0$ cal
(1)	2.0	863	gro + qz	$\Delta H^0_{1\ bar}$	$- 863\Delta S^0$	$> - 1561$
(2)	2.0	883	an + woll	$\Delta H^0_{1\ bar}$	$- 883\Delta S^0$	$< - 1561$
(3)	3.0	893	gro + qz	$\Delta H^0_{1\ bar}$	$- 893\Delta S^0$	$> - 2342$
(4)	3.0	923	an + woll	$\Delta H^0_{1\ bar}$	$- 923\Delta S^0$	$< - 2342$
(5)	4.7	973	gro + qz	$\Delta H^0_{1\ bar}$	$- 973\Delta S^0$	$> - 3669$
(6)	4.7	993	an + woll	$\Delta H^0_{1\ bar}$	$- 993\Delta S^0$	$< - 3669$
(7)	6.0	1023	gro + qz	$\Delta H^0_{1\ bar}$	$- 1023\Delta S^0$	$> - 4684$
(8)	5.5	1023	an + woll	$\Delta H^0_{1\ bar}$	$- 1023\Delta S^0$	$< - 4294$

$\Delta V^0 = + 0.7806$ cal bar^{-1} (32.66 cm^3)

The 1 in (P-1) ΔV^0 has been neglected.

of this line. This approach is illustrated in Fig. 7.2 for experiment number (1) in Table 7.1. The line corresponding to eqn (7.23) has been constructed by using T = 863, $P\Delta V^0 = 1561$, and two values of ΔS^0, + 15 e.u. and + 25 e.u.

For each value of ΔS^0 a <u>minimum</u> value of $\Delta H^0_{1\ bar}$ consistent with the experimental result is obtained. The estimate of $\Delta H^0_{1\ bar}$ is a minimum because for this run

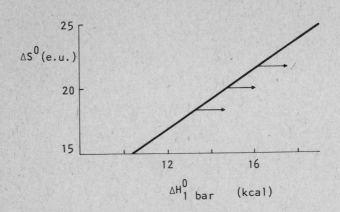

FIG. 7.2.   The line $\Delta H^0_{1\ bar} - 863\ \Delta S^0 = -1561$ plotted for experiment (1) in ΔH^0, ΔS^0 space.   Since for this experiment $\Delta H^0_{1\ bar} - 863 \Delta S^0 > -1561$, the internally consistent values of $\Delta H^0_{1\ bar}$, ΔS^0 lie to the right of the line.

eqn (7.21) (reactants stable) applies. This same procedure has been followed for each experiment described in Table 7.1 and the result is illustrated in Fig. 7.3. Bearing in mind the result of each experiment, the intersection of lines leads to a very small region of possible values of $\Delta H^0_{1\ bar}$, ΔS^0 which are consistent with all the data. This region, shaded in Fig. 7.3, corresponds to a ΔS^0 of reaction of between 17.1 e.u. and 22.1 e.u. and $\Delta H^0_{1\ bar}$ of 13.2 to 18.0 kcal. Note that the values of ΔS^0 and ΔH^0 are interdependent. If we were to adopt the highest possible figure for ΔS^0

(22.1 e.u.) this would only be consistent with ΔH^0 of 18.0 kcal. Similarly a ΔS^0 of 17.1 e.u. is consistent only with a $\Delta H^0_{1\ bar}$ of + 13.2 kcal.

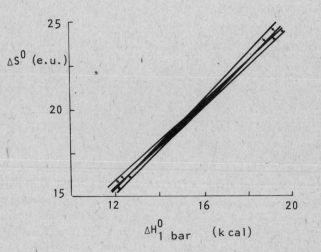

FIG. 7.3. The intersection of lines in $\Delta H^0_{1\ bar}$ - ΔS^0 space for the experimental results in Table 7.1. The shaded area is the region of possible values of $\Delta H^0_{1\ bar}$ and ΔS^0 which are consistent with all the experimental data (some lines omitted for clarity). The ranges of values are : $\Delta H^0_{1\ bar}$, 13.2 to 18.0 kcal; ΔS^0, 17.1 to 22.1 e.u.

It is now necessary to decide which values of $\Delta H^0_{1\ bar}$ and ΔS^0 to adopt from the maximum possible range derived above. The most reasonable approach is to assume that the 'best'

estimate of ΔS^0 is that value which gives the maximum
range of internally consistent values of $\Delta H^0_{1\,bar}$ (equivalent
to maximum uncertainty). This is equivalent to fitting a
'best' slope to the experimental brackets, i.e. a slope
which most closely bisects all of the bracketing points
shown in Fig. 7.1. The 'best' estimate of ΔS^0 is usually
close to halfway between the two extreme values; in this
case 19.6 e.u. Before adopting this value, however, it is
necessary to confirm that a ΔS^0 of 19.6 e.u. is reasonably
consistent with the tabulated entropy values in Robie and
Waldbaum (1968). At 298 K the ΔS^0 of reaction (7.21)
taken from Robie and Waldbaum is 20.07 ± 1.4 e.u.; at
900 K it is 19.53 e.u. (with similar uncertainty). The
uncertainty estimates have been obtained by combining errors
in the normal way :

$$\sigma_{\Delta S} = \left[(\sigma_{gro})^2 + (\sigma_{qtz})^2 + (\sigma_{an})^2 + 2\,(\sigma_{woll})^2 \right]^{1/2} \quad (7.24)$$

where σ_{gro} etc. are the tabulated uncertainties in the
entropies of the subscript phases and $\sigma_{\Delta S}$ is the uncertainty
in ΔS^0. The reason for the relatively large uncertainty in
the tabulated ΔS^0 is because the entropy of grossular was
derived from experimental data and the heat capacity
adopted for grossular is an estimate of the type discussed
in Chapter 1. It is apparent that a value of ΔS^0 of
+ 19.6 e.u. is in excellent agreement with other calorimetric
and experimental data and may be adopted with a reasonable
degree of confidence. Substituting ΔS^0 of 19.6 e.u. into

each of the inequalities of Table 7.1 indicates that a $\Delta H^0_{1\ bar}$ of between 15 402 cal and 15 746 cal is consistent with the experimental results. The final expression for $\Delta G^0_{P,T}$ of reaction (7.21) is therefore

$$\Delta G^0_{P,T} = 15\ 574 - 19.6\ T + (P-1)\ 0.7806 \pm 172\ cal \qquad (7.25)$$
$$= (65\ 162 - 82.0\ T + (P-1)\ 3.266 \pm 720\ J).$$

This may now be applied to natural assemblages provided compositions of phases and activity-composition relationships are known. It should be noted that the estimated uncertainty in $\Delta G^0_{P,T}$, $\pm$ 172 cal, is equivalent to very small errors in temperature and pressure, $\pm\ 9^oC$ or $\pm$ 220 bar respectively. The reason for the small size of the uncertainty is that the experimental points bracket the equilibrium boundary very closely ($\pm\ 10^oC$) at several pressures. The actual error in $\Delta G^0_{P,T}$ may, however, be somewhat larger than that quoted because of uncertainties in the pressures and temperatures at which the experiments were performed. We have assumed in this treatment that the experimental pressures and temperatures quoted by Newton (1966) and Boettcher (1970) are correct, whereas in practice there are small uncertainties associated with the measurement of pressure and temperature. There are various ways of taking account of possible experimental uncertainties in the errors and one way which yields maximum likely errors is discussed in section 7.6.

7.5   REACTIONS INVOLVING VOLATILES

A number of different procedures for extracting thermodynamic data from experiments involving volatiles have
appeared in the literature. The best method in any
particular case depends on the use which is to be made of
the data. For the simplest case of obtaining thermodynamic
data for the reaction of interest and using only small
temperature extrapolations, it is reasonable to assume
$\Delta C_p = 0$ and to use an analogous approach to that described
for solid-solid reactions. If a large body of heat of
formation data is to be obtained then it is necessary to
use the available heat capacities in order to obtain
accurate results for the reference conditions of 298 K
and 1 bar.

Consider the reaction

$$Mg_7Si_8O_{22}(OH)_2 \rightleftharpoons 3\tfrac{1}{2}Mg_2Si_2O_6 + SiO_2 + H_2O, \qquad (7.26)$$
$$\text{anthophyllite} \qquad \text{enstatite} \qquad \text{quartz} \qquad \text{fluid}$$

which has been studied by Greenwood (1963). Greenwood's
experimental data, which include only one well-defined
'bracket' (at 2 kbar, $P_{H_2O} = P_{total}$), are summarized in
Table 7.2.

At equilibrium for reaction (7.26) we have

$$\mu^{anth}_{Mg_2Si_8O_{22}(OH)_2} = 3\tfrac{1}{2}\mu^{enst}_{Mg_2Si_2O_6} + \mu^{qz}_{SiO_2} + \mu^{fluid}_{H_2O}. \qquad (7.27)$$

Taking standard states of solid components to be the pure

TABLE 7.2

Experimental results for the reaction

anthophyllite $\rightleftharpoons$ enstatite + quartz + H_2O

from Greenwood (1963)

P_{H_2O}(kbar)	$T(^oC)$	Stable Assemblage	
1.0	754	enst + qz + H_2O	(a)
2.0	760	anthophyllite	(b)
2.0	770	enst + qz + H_2O	(c)
2.6	750	anthophyllite	(d)
2.6	775	enst + qz + H_2O	(e)

phases at the pressure and temperature of interest, and the standard state of H_2O to be pure H_2O at 1 bar and the temperature of interest yields

$$(\mu^0_{Mg_2Si_8O_{22}(OH)_2})_{P,T} = 3\tfrac{1}{2}\,(\mu^0_{Mg_2Si_2O_6})_{P,T} + (\mu^0_{SiO_2})_{P,T} +$$

$$+ (\mu^0_{H_2O})_{1,T} + RT \ln a_{H_2O}. \qquad (7.28)$$

Rearranging (7.28), which is applicable to Greenwood's experiments because all phases were pure, and noting that the activity of H_2O in a one-component fluid is equal to the fugacity of H_2O, yields

$$\Delta G^0 = - RT \ln f_{H_2O} \tag{7.29}$$

and

$$\Delta H^0_{1 \text{ bar},T} - T\Delta S^0_T + (P-1) \Delta V^0_{\text{solids}} = - RT \ln f_{H_2O}. \tag{7.30}$$

A simple treatment of the data involves assuming that $\Delta H^0_{1 \text{ bar}}$ and ΔS^0 are constant in the temperature range of interest and obtaining thermodynamic data in the manner already described for the grossular-quartz reaction, i.e.

$$\Delta H^0_{1 \text{ bar}} - T\Delta S^0 > - RT \ln f_{H_2O} - (P-1) \Delta V^0_{\text{solids}} \tag{7.31}$$
$$\text{(reactants stable)}$$

or

$$\Delta H^0_{1 \text{ bar}} - T\Delta S^0 < - RT \ln f_{H_2O} - (P-1) \Delta V^0_{\text{solids}} \tag{7.32}$$
$$\text{(products stable)}$$

where

$$\Delta V^0_{\text{solids}} = - 22.09 \text{ cm}^3 = - 0.528 \text{ cal bar}^{-1}.$$

Graphical solution of eqns (7.31) and (7.32) using $f_{H_2O} - P$ data from Burnham et al. (1969) gives, as was pointed out by Greenwood, an extremely wide range of internally-consistent values of $\Delta H^0_{1 \text{ bar}}$ and ΔS^0. The enthalpy change may be almost any value greater than 10 kcal and ΔS^0 any value greater than about 25 e.u. The lack of constraint on $\Delta H^0_{1 \text{ bar}}$ and ΔS^0 arises from the fact that only one closely bracketed pressure-temperature point was determined by

Greenwood. Two or more close brackets are needed in order to obtain tightly constrained values of $\Delta H^0_{1 \, bar}$ and ΔS^0.

In the absence of well-defined values of entropy and enthalpy of reaction (7.26), it is desirable to fix ΔS^0 as closely as possible and to derive the enthalpy of reaction on the basis of this assumed ΔS^0. This procedure is usually reasonable because even in the absence of entropy data for solid phases, it is generally possible to make adequate entropy estimates. The entropies of enstatite (clino-), quartz, and all H_2O modifications are tabulated in Robie and Waldbaum (1968). The entropy of anthophyllite has not been measured and it is proposed to estimate it from the oxide values or from appropriate simple silicates. Initially, $(S^0_{298})_{anthophyllite}$ was estimated directly from oxides ($7MgO + 8(SiO_2)_{quartz} + (H_2O)_{ice}$) with a correction for the difference in volumes of oxides and silicate (eqn (1.53)). The value obtained is about 130 e.u. Because of the relatively large correction for volume ($\Delta V = -15.4 \, cm^3$), however, the uncertainty in this value is quite large. Alternatively an entropy estimate on the basis of silicates with similar structure to anthophyllite would be expected to yield a more reliable result. To make this estimate we may use the values of clinoenstatite (a chain silicate), quartz, and ice with the following results :

$$7(S^0_{298})_{clinoenst} + (S^0_{298})_{qtz} + (S^0_{298})_{ice} = 134.1 \text{ e.u.}$$

With volume correction,

$$(S^0_{298})_{anthophyllite} = 135.0 \text{ e.u.}$$

The entropy of anthophyllite at 298.15 K is estimated to be 135.0 e.u., a figure which is in good agreement with the values adopted by Greenwood (1963) (136.6 e.u.) and Zen (1971) (133.62 e.u.). At 298 K the entropy change of reaction (7.26), with water in the standard state of steam at 1 bar, is + 33.5 e.u. The entropy change at the temperature of Greenwood's experiments (approximately 1038 K) may be obtained by integrating $(\Delta C_p/T)$ dT between 298 and 1038 K. In the cases of clinoenstatite (entropies assumed applicable for orthoenstatite), quartz, and steam the entropies at high temperature are readily derived from the tables of Robie and Waldbaum by interpolation between the values given at 1000 K and 1100 K. For anthophyllite, S^0_{1038} was obtained in the normal way from the heat-capacity function (Birch <u>et al</u>. 1942) :

$$(C_p)_{anth} = 199 + 34.1 \times 10^{-3} \ \frac{- 52.3 \times 10^5}{T^2} \ \text{cal mol}^{-1}\text{deg}^{-1}. \tag{7.33}$$

Combining the data yields ΔS_{1038} of + 33.3 e.u. In this case the change in ΔS^0 between 298 K and 1038 K is extremely small; many other dehydration reactions exhibit much larger changes in ΔS^0 with changing temperature.

Assuming that ΔS^0 is equal to 33.3 e.u. and constant within the temperature range of interest, the value of $\Delta H^0_{1 \text{ bar}}$ obtained from Greenwood's experiments must lie between + 20 454 and + 20 477 cal. From the assumption of $\Delta C_p = 0$,

the standard-state free energy change of reaction (7.26) is
given by

$$\Delta G^0 = 20\ 465 - 33.3T + (P-1)\ 0.528 = -RT\ \ln K\ (cal)$$
$$= (85\ 626 - 139.3T + (P-1)\ 2.209 = -RT\ \ln K\ (J). \quad (7.34)$$

Equations of the form (7.34) are often expressed in terms
of $\log_{10} K$, i.e.

$$\log_{10} K = \frac{-\ 20\ 465}{2.303\ RT} + \frac{33.3}{2.303R} + \frac{(P-1)\ 0.528}{2.303RT} \quad (7.35)$$

or

$$\log_{10} K = \frac{4470}{T} + 7.28 + \frac{(P-1)\ 0.1154}{T}. \quad (7.36)$$

Expressions (7.34), (7.35), or (7.36) may now be used to
calculate equilibria involving anthophyllite, enstatite,
quartz, and vapour, provided that the temperatures involved
do not differ greatly from the temperatures at which
Greenwood's experiments were performed. The errors involved
in assuming that $\Delta H^0_{1\ bar}$ and ΔS^0 are constant will be
discussed in more detail later in this section.

In order to obtain enthalpy and entropy data for reaction
(7.26) which are applicable over wide temperature ranges it
is, of course, necessary to take account of the heat
capacities of solid and fluid phases.

The heat capacities of enstatite (using clinoenstatite
data), quartz, and H_2O (steam) at 1 bar are as follows :

clinoenstatite : $C_p = 24.55 + 4.74 \times 10^{-3}T$

$$\frac{- 6.28 \times 10^5}{T^2} \quad \text{cal mol}^{-1} \text{ deg}^{-1}$$

α-quartz $\qquad$: $C_p = 11.22 + 8.2 \times 10^{-3}T$

$$\frac{- 2.7 \times 10^5}{T^2} \quad \text{cal mol}^{-1} \text{deg}^{-1} \tag{7.37}$$

β-quartz $\qquad$: $C_p = 14.41 + 1.94 \times 10^{-3}T$ cal mol^{-1} deg^{-1}

steam $\qquad$: $C_p = 7.30 + 2.46 \times 10^{-3}T$ cal mol^{-1} deg^{-1}

These data give values of ΔC_p for reaction (7.26) as follows :

α-quartz field : $\Delta C_p^{\alpha\text{-qz}} = - 8.63 + 9.74 \times 10^{-3}T +$

$$+ \frac{5.64 \times 10^{-5}}{T^2} \text{ cal mol}^{-1} \text{ deg}^{-1} \tag{7.38}$$

β-quartz field : $\Delta C_p^{\beta\text{-qz}} = - 5.44 + 3.48 \times 10^{-3}T +$

$$+ \frac{8.34 \times 10^{-5}}{T^2} \text{ cal mol}^{-1} \text{ deg}^{-1} \tag{7.39}$$

Taking account of the enthalpy and entropy of the reaction

$$\begin{array}{cc} SiO_2 & \rightleftarrows & SiO_2 \\ \alpha\text{-quartz} & \beta\text{-quartz} \end{array} \tag{7.40}$$

(equilibrium temperature at 1 bar = 838 K, standard state $\Delta H_{1\ bar,848} = 0.29$ kcal, $\Delta S^0 = 0.34$ e.u.), the free energy change of reaction (7.26) at 1 bar and T is given by (in the α-quartz field) :

$$\Delta G^0_{1,T} = \Delta H^0_{1\ bar,298} + \int_{298}^{T} \Delta C_p^{\alpha\text{-}qz}\ dT - T \left(\Delta S^0_{298} + \right.$$

$$\left. + \int_{298}^{T} \frac{\Delta C_p^{\alpha\text{-}qz}}{T}\ dT \right) \text{cal.}$$

(7.41)

In the β-quartz field, $\Delta G^0_{1,T}$ is given by

$$\Delta G^0_{1,T} = \Delta H^0_{1\ bar,298} + \int_{298}^{848} \Delta C_p^{\alpha\text{-}qz}\ dT + 290 + \int_{848}^{T} \Delta C_p^{\beta\text{-}qz}\ dT - $$

$$- T \left(\Delta S^0_{298} + \int_{298}^{848} \frac{\Delta C_p^{\alpha\text{-}qz}}{T}\ dT + 0.34 + \int_{848}^{T} \frac{\Delta C_p^{\beta\text{-}qz}}{T}\ dT \right) \text{cal.}$$

(7.42)

Since all of Greenwood's experiments were performed in the β-quartz field, eqn (7.42) is the one which must be used to extract a value of $\Delta H^0_{1,298}$. The left-hand sides of eqns (7.31) and (7.32) are both $\Delta G^0_{1,T}$, and so the experiments may be used to determine the range of possible values of $\Delta G^0_{1,T}$ at the experimental temperatures. Hence, from (7.42) the possible values of $\Delta H^0_{1,298}$ may be derived. Applying inequalities (7.31) and (7.32) and eqn (7.42) to the results in Table 7.2 yields the following values of $\Delta G^0_{1,T}$ and corresponding inequalities for $\Delta H^0_{1,298}$:

(a) $\Delta G^0_{1,1027} < -13\ 148$ cal $\Delta H^0_{1,298} < 21\ 296$ cal

(b) $\Delta G^0_{1,1033} > -13\ 945$ cal $\Delta H^0_{1,298} > 20\ 704$ cal

(c) $\Delta G^0_{1,1043} < -14\ 117$ cal $\Delta H^0_{1,298} < 20\ 842$ cal

(d) $\Delta G^0_{1,1023} > -13\ 965$ cal $\Delta H^0_{1,298} > 20\ 351$ cal

(e) $\Delta G^0_{1,1048} < -14\ 421$ cal $\Delta H^0_{1,298} < 20\ 731$ cal

(Runs denoted by letters as above and in Table 7.2.)

The range of values of $\Delta H^0_{1,298}$ which is compatible with the experimental data is $20\ 717 \pm 14$ cal $(86\ 680 \pm 59$ J). Although it is extremely arduous to calculate enthalpy data using eqn (7.42) by hand, it is comparatively simple to write a computer program to extract thermodynamic data using this approach. The 1 bar/298 enthalpy of reaction (7.26) derived above may now be substituted back into eqn (7.42) to calculate $\Delta G^0_{1,T}$, and hence equilibria at any pressure and temperature, using the standard states of 1 bar and T for H_2O and P bar and T for solids :

$$\Delta G^0_{1,T} + (P-1)\ \Delta V^0_{solids} = - RT \ln K. \tag{7.43}$$

The use of heat-capacity functions for solid and fluid phases leads, of course, to greater accuracy in $\Delta G^0_{1,T}$ outside the temperature range of the experiments than does the assumption of $\Delta C_p = 0$. This additional accuracy may or may not be necessary in any particular case, and before deciding

on which approach to take it is desirable to consider the
magnitudes of errors occurring from sources other than the
thermodynamic data. If, for example, activity-composition
relationships of complex phases are only poorly known, then
the errors introduced by assuming $\Delta C_p = 0$ may be much
smaller than possible errors in component activities. In
that case the use of eqns (7.34) to (7.36) would be adequate.
It is of interest to note that for reaction (7.26), extrapolation of free energy data from 1038 K to 298 K assuming
$\Delta C_p = 0$ introduces an error in $\Delta G^0_{1,298}$ of only 200 cal or
less than 6°C in calculated equilibrium temperature. Clearly,
in this case the simple approach is adequate for petrological
calculations.

Given a value of $\Delta H^0_{1,298}$ for reaction (7.26), the heat of
formation of any of the phases from its constituent
elements may be derived provided that all of the others are
known. By analogy with (7.10) we have :

$$\Delta H^0_{1,298} = 7(H^0_f)_{enst} + (H^0_f)_{qz} + (H^0_f)_{steam} - (H^0_f)_{anthoph}$$

$$\text{(7.44)}$$

$$= 20\ 717\ cal\ (= 86\ 680\ J).$$

Using the H^0_f values of (clino-) enstatite, quartz, and steam
given in Robie and Waldbaum, (1968) the enthalpy of formation of
anthophyllite from its constituent elements is

$$(H^0_f)_{anthoph\ 298,1} = 7 \times (-\ 370\ 140) - 217\ 650 - 57\ 796 - 20\ 717$$

$$= -\ \underline{2\ 887\ 143}\ cal\ mol^{-1}$$

$$\text{(7.45)}$$

$$(= -\ 12\ 079.8\ kJ\ mol^{-1}).$$

(Note that, following Zen (1971), we have ignored the enthalpy (unknown) of the ortho-clino inversion in enstatite.)

A slightly different approach to obtaining enthalpy and free energy of formation data from dehydration reactions (Fisher and Zen 1971) uses the function $G^*_{H_2O}$:

$$(G_{f,H_2O})_{1,T} + \int_1^P V_{H_2O}\ dP = G^*_{H_2O}. \qquad (7.46)$$

In eqn (7.46), $(G_{f,H_2O})_{1,T}$ is the free energy of formation of steam from hydrogen and oxygen at (1 bar and T). Fisher and Zen combine $G^*_{H_2O}$ (which they tabulate) with the assumption that the differences between the entropies of formation of solid phases from the elements are constant. In this case

$$\Delta S_{f,s} = 7S_{f,en} + S_{f,qz} - S_{f,anth} = const. \qquad (7.47)$$

which gives, for any P and T,

$$(\Delta G^0)_{P,T} = 7(G_{f,en})_{1,298} + (G_{f,qz})_{1,298} - (G_{f,anth})_{1,298} +$$
$$\qquad\qquad\qquad\qquad\qquad\qquad\qquad\qquad\qquad\qquad\qquad (7.48)$$
$$+ (P-1)\ \Delta V_{solids} + (T-298)\ \Delta S_{f,solids} + G^*_{H_2O}.$$

The free energy of formation (and hence enthalpy of formation) at 1 bar, 298 K of any one of the three phases enstatite, quartz, and anthophyllite may now be obtained provided that the other two are known :

$$(G_f)_{1,298} = (H_f)_{1,298} - 298 \, S_f. \tag{7.49}$$

When using tabulated H_f, G_f data, the reader should make
sure which, if any, assumptions have been made in its
derivation. In order to generate the correct values of
$(\Delta G^0)_{P,T}$ it is necessary to use the same set of assumptions
($\Delta C_p = 0$, $\Delta S_{f,\text{solids}} = $ const. etc.) used by the authors in
deriving the data set.

7.6 EXPERIMENTAL ERRORS

In discussing the procedure for extracting thermodynamic
data it has been assumed that the pressures and temperatures
quoted by the experimentalists are always correct. There are,
however, uncertainties attached to the pressures and temperatures of experimental runs and it is necessary to consider
how these affect the possible errors in derived thermodynamic
data.

Approximate uncertainties (conservative) in pressure
and temperature are given in Table 7.3 for the different
types of apparatus which are generally used in experimental
studies. As can be seen, the possible errors in pressure
(taken from Gordon 1973) are much lower if pressure is
applied hydrostatically via a fluid medium (hydrothermal and
internally-heated vessels) than if the pressure medium is a
deformable solid. Temperature uncertainties depend, of
course, on the magnitude of temperature gradients in the
apparatus concerned, the cold-seal externally-heated type
generally having the lowest temperature gradient and the
greatest precision of temperature measurement.

TABLE 7.3

Approximate uncertainties in experimental pressures
and temperatures (conservative)

Apparatus	Pressure uncertainty	Temperature uncertainty
Hydrothermal cold seal	$\pm$ 1%	$\pm$ 5°C
Internally heated	$\pm$ 1%	$\pm$ 5-10°C
Solid media piston-cylinder	$\pm$ 0.5-1 kb	$\pm$ 10-15°C

There is no well-defined method for combining the
uncertainties in experimental run conditions with the
uncertainties in internally-consistent values of $\Delta H^0_{1 \text{ bar}, T}$
and ΔS^0_T. Exactly how this is done depends on whether one
wants to make an estimate of likely error in calculated
pressures and temperatures or whether an estimate of maximum
possible error is required. If an estimate of 'likely' error
is adequate then it is probably reasonable to use the range
of ΔH^0 and ΔS^0 values derived by assuming run pressures and
temperatures to be correct. If maximum error estimates are
desired, then the procedure adopted by Gordon (1973) is
appropriate. The method of Gordon is to assume that the actual
pressure and temperature of any run, as opposed to the quoted
values, corresponded to the pressure and temperature furthest
away from the equilibrium boundary which is consistent with
the experimental uncertainties.   This is illustrated
in Fig. 7.4 for some reaction A $\rightleftarrows$ B.   The points

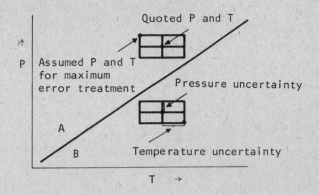

FIG. 7.4. A comparison of intended experimental pressures
and temperatures with those assumed for a maximum error
treatment.

used to calculate the range of internally-consistent
enthalpy and entropy values are those in the corners of the
uncertainty 'boxes' furthest away from the equilibrium
boundary. This approach gives the maximum possible range
of internally-consistent $\Delta H^0_{1\ bar,T}$, ΔS^0_T values.

In attempting to derive thermodynamic data from experiments, it will be found in some cases that there are
several sets of experimental results on the same reaction
published by different authors. These data sets are often
internally consistent but not consistent with one another.

Examples of frequently studied reactions with external
inconsistencies are

$$KAl_3Si_3O_{10}(OH)_2 + SiO_2 \rightleftarrows KAlSi_3O_8 + Al_2SiO_5 + H_2O$$

muscovite	quartz	feldspar	andalusite	fluid
			or	
			sillimanite	

(Evans 1965, Althaus et al. 1970, Day 1973, Chatterjee and Johannes 1974) and the Al_2SiO_5 phase diagram (Bell 1963, Weill 1966, Richardson et al. 1969, Holdaway 1971).

Inconsistencies arise between different sets of experiments because of variations in the structural state of starting materials (e.g. Al-Si disorder in feldspar and sillimanite), because of production of metastable assemblages which do not react and, of course, because of errors in pressure and temperature. If there are two or more inconsistent sets of data it is necessary to judge which of the experiments are likely to be the most accurate and to discard unsatisfactory results.

The criteria to apply when considering the relative merits of different experimental results are as follows :

(1)   The structural states and compositions of all phases involved in the reaction should be well characterized both before and, if possible, after the experiment. This is to ensure that the experimental results definitely apply to the reaction which the experimentalist is attempting to study.

(2)   Starting material should be mixtures of crystalline product and reactant phases together with the appropriate fluid phase. Preferably no amorphous (glass or gel) starting material should be used. The

reason for this is that glasses and gels have in
most cases much higher free energies than either
product or reactant phases involved in the reaction
of interest. Since they are unstable with respect
to both products and reactants, they may crystallize
initially to either the stable or the unstable phases.
If the unstable phases crystallize from the amorphous
material first, it is possible that they will not
react during the duration of the experiment and hence
the wrong result may be obtained. In general,
amorphous materials crystallize initially to the
phases with the highest entropy rather than those
with the least free energy.

(3)   Only experiments which 'bracket' the equilibrium
boundary should be used. The fact that no reaction
took place does not necessarily mean that the run
was close to the equilibrium boundary.

(4)   For each bracketing run a reasonable amount of
reaction must be shown to have taken place. The
experimentalist usually determines the direction of
reaction from a combination of optical and, if the
amount of reaction is low, X-ray techniques. If the
direction of reaction is determined from the relative
peak heights on X-ray diffractograms, a large change
(40 percent or so) in relative peak heights of
reactants and products should have taken place. This
is because small variations in sample preparation,

grinding, and mounting can have considerable
effects on relative peak heights.

(5)   In general, it is desirable to have starting
materials, as fine-grained as possible, because
the greater the surface area of reactants, the
more rapidly the reaction takes place. It has been
shown, however, that if the entropy and volume
changes of the reaction are small, excessive
grinding may shift the apparent position of the
equilibrium boundary (Newton 1969). This is
because the molar free energy of phases increases
as the surface area to volume ratio increases
(grain size decreases), and in addition because
grinding produces dislocations in solids which add
to their free energies. Although the absolute
effect of grinding the material is likely to be
only of the order of a few tens of calories on
the free energy of a reaction, the result could
be to shift the boundary considerably in P-T space
if ΔS and ΔV are small.

Experimental data sets should, ideally, match up to all
five of these criteria. In practice, of course, this is not
so common and it is necessary to weight the usefulness of
individual experiments on the basis of how closely they
approach the ideal. It is hoped that this section will enable
the reader to make reasonable judgements of the relative
merits of data sets which are externally inconsistent.

SUMMARY - CHAPTER 7

1. The enthalpies and free energies of many pure phases
    of geological interest have been determined relative to
    their constituent elements by calorimetric measurements.
    These data may, in principle, be used to determine the
    standard-state free energy change (pure phases at Pbar
    and T) of any reaction of interest.

2. There are, in some cases, internal inconsistencies in
    calorimetric data sets which have arisen because of
    errors in the enthalpy data for standard (reference)
    compounds. The inconsistency of the value of (H_f) for
    $Al(OH)_3$ gibbsite relative to H_F for $\alpha-Al_2O_3$ has caused
    considerable difficulty.

3. An alternative approach to the determination of
    standard-state free energy data is to extract it directly
    from experimental results which 'bracket' the reaction
    curve of interest, e.g.

$$2A + B \rightleftarrows C \qquad\qquad \Delta H_1^0.$$

    Experiments in which pure C was observed to form from
    pure A and pure B have the condition (taking standard
    states to be pure components at the P and T of interest) :

$$\Delta G_{P,T} = 0.$$

$$= \Delta G_{P,T}^0 = \Delta H_{1,T}^0 - T\Delta S_T^0 + \int_1^P \Delta V^0 \, dP < 0.$$

If the reactants (A and B) are stable relative to C, then we have

$$\Delta G^0_{P,T} = \Delta H^0_{1,T} - T\Delta S^0_T + \int_1^P \Delta V^0 \, dP > 0.$$

Setting up inequalities of this type for runs on either side of the equilibrium boundary enables one to determine the range of values of $\Delta H^0_{1,T}$ and ΔS^0_T which are consistent with the experimental data. Assuming that ΔV^0_{solids} is constant, we have

$$\int_1^P \Delta V^0 \, dP = (P-1) \, \Delta V^0,$$

and for a solid-solid reaction

$$\Delta G^0_{P,T} = \Delta H^0_{1,T} - T\Delta S^0_T + (P-1) \, \Delta V^0 \lessgtr 0.$$

This approach may be extended to reactions involving fluids using, for example, a standard state of 1 bar and T for the fluid components. In the case of reactions involving solids only, it is reasonable to assume $\Delta C_p = 0$. Reactions involving fluids may also be treated in this way over small temperature ranges, although heat capacity values should be used if large temperature extrapolations are necessary.

4.   If experimental data are not available for the reaction of interest, results for related reactions may be combined to

derive the required enthalpy and entropy data, e.g.

Required reaction : $2A + D \rightleftarrows C$ $\Delta H^0_2, \quad \Delta S^0_2$

Available data : $2A + B \rightleftarrows C$ $\Delta H^0_1, \quad \Delta S^0_1$

                   $B \rightleftarrows D$ $\Delta H^0_3, \quad \Delta S^0_3$

$$\Delta H^0_2 = \Delta H^0_1 - \Delta H^0_3; \qquad \Delta S^0_2 = \Delta S^0_1 - \Delta S^0_3.$$

5. Given enthalpy changes derived from experiments, heat of formation data for any one of the phases involved in the reaction may be obtained provided values for the other phases are known, e.g.

$$(\Delta H^0_1)_T = (H^C_T - H^B_T - 2H^A_T) = (H^C_f)_T - (H^B_f)_T - 2(H^A_f)_T.$$

6. Where external inconsistencies arise between different sets of experimental data on (nominally) the same reaction, it is necessary to make some judgement as to which results are more likely to be correct. The relevant criteria were described in section 7.6.

PROBLEMS - CHAPTER 7

1. (a) The reaction

                                                       (1)

$$KAl_3Si_3O_{10}(OH)_2 + SiO_2 \rightleftarrows KAlSi_3O_8 + Al_2SiO_5 + H_2O$$
     muscovite   quartz   sanidine   andalusite   fluid

has been studied by Chatterjee and Johannes (1974).
The following results were obtained

	$P_{H_2O} = P_{total}$ (kbar)	$T(^oC)$	f_{H_2O} (bar)	Result
(1)	0.5	560	368	fsp + and increased
(2)	0.5	540	355	No reaction
(3)	0.5	520	339	musc + qz increased
(4)	1.0	570	584	fsp + and increased
(5)	1.0	550	550	musc + qz increased
(6)	2.0	605	1062	fsp + and increased
(7)	2.0	595	1030	No reaction
(8)	2.0	590	1014	musc + qz increased
(9)	3.0	640	1722	fsp + and increased
(10)	3.0	630	1676	No reaction
(11)	3.0	620	1628	musc + qz increased
(12)	4.0	670	2613	fsp + and increased
(13)	4.0	660	2551	musc + qz increased
(14)	5.0	705	3849	fsp + and increased
(15)	5.0	690	3727	musc + qz increased

Standard state of H_2O is pure H_2O at 1 bar and T^oC.
Determine the range of internally-consistent values of
ΔS_T^0, $\Delta H_{1,T}^0$ assuming that run pressures and temperatures
are correct and $\Delta C_p = 0$ in the temperature range of
interest, and

$$\Delta V^0_{solids} = - 2.82 \ cm^3 = - 0.0674 \ cal \ bar^{-1}$$

(assume constant)

(b) Third law entropies in e.u. (Robie and Waldbaum 1968).

	H_2O (steam)	Sanidine	Andalusite	Quartz	Muscovite
$S_{1 \ bar, 298}$	45.10 (0.01)	56.94 (1.0)	22.28 (0.1)	9.88 (0.02)	69.00 (0.1)
$S_{1 \ bar, 800}$	53.46	118.05	60.08	23.76	166.24
$S_{1 \ bar, 900}$	54.57	126.76	65.39	26.09	179.91
$S_{1 \ bar, 1000}$	55.59	134.66	70.23	27.80	192.52

T is in degrees K. The value in brackets is the uncertainty in the 298 K entropy.

Adopt the most reasonable value of $\Delta S^0_{1,T}$ in the temperature range of interest and calculate maximum and minimum values of $\Delta H^0_{1,T}$ consistent with this entropy change.

2. Calculate the heat of formation of muscovite at 298 K 1 bar $(H^{musc}_f)_{1,298}$ using the following data from Robie and Waldbaum (1968) and $\Delta H^0_{1,T}$ (derived from question (2)) :

	$(H_f)_{1,298}$ cal mol^{-1}	$(H_{900} - H_{298})$ (cal)
Quartz	$-$ 217 650	9300
Muscovite	$-$	62 529
Sanidine	$-$ 944 378	39 400
Andalusite	$-$ 619 390	24 320
Steam	$-$ 57 796	5240

SOLUTIONS TO PROBLEMS – CHAPTER 7

1. Neglecting experiments on the reaction

$$KAl_3Si_3O_{10}(OH)_2 + SiO_2 \rightleftarrows KAlSi_3O_8 + Al_2SiO_5 + H_2O$$
$$\text{muscovite} \qquad \text{quartz} \quad \text{sanidine} \quad \text{andalusite} \quad \text{fluid}$$

in which no change was observed – numbers (2), (7) and (10) – and assuming ΔV^0_{solids} constant and $\Delta C_p = 0$, the appropriate expressions are :

$$\Delta H^0_{1,T} - T\Delta S^0_T < - RT \ln f_{H_2O} - (P-1)\, \Delta V^0_{solids}$$

(sanidine and andalusite stable)

$$\Delta H^0_{1,T} - T\Delta S^0_T > - RT \ln f_{H_2O} - (P-1)\, \Delta V^0_{solids}.$$

(muscovite and quartz stable)

Substituting the values of T, P ΔV^0_{solids} and f_{H_2O} given :

Run No.

(1) $H^0_{1,T} - 833 \Delta S^0_T < - 9745$ cal

(3) $H^0_{1,T} - 793 \Delta S^0_T > - 9146$ cal

(4) $H^0_{1,T} - 843 \Delta S^0_T < - 10\ 603$ cal

(5) $H^0_{1,T} - 823 \Delta S^0_T > - 10\ 252$ cal

(6) $H^0_{1,T} - 878 \Delta S^0_T < - 12\ 021$ cal

(8) $H^0_{1,T} - 863 \Delta S^0_T > - 11\ 734$ cal

(9) $H^0_{1,T} - 913 \Delta S^0_T < - 13\ 316$ cal

(11) $H^0_{1,T} - 393 \Delta S^0_T > - 12\ 920$ cal

(12) $H^0_{1,T} - 943 \Delta S^0_T < - 14\ 473$ cal

(13) $H^0_{1,T} - 933 \Delta S^0_T > - 14\ 272$ cal

(14) $H^0_{1,T} - 978 \Delta S^0_T < - 15\ 706$ cal

(15) $H^0_{1,T} - 963 \Delta S^0_T > - 15\ 398$ cal

Plotting ΔS^0_T versus $\Delta H^0_{1,T}$ as before, the approximate ranges of internally-consistent ΔS^0_T, $\Delta H^0_{1,T}$ values are :

ΔH^0_T = + 35 to + 40 e.u.

$\Delta H^0_{1,T}$ = + 18 to + 23 kcal.

The 'best-fit' value of ΔS^0 which corresponds to the widest possible range of consistent $\Delta H^0_{1,T}$ values is about

+ 37.5 e.u.   This figure is slightly smaller than those
calculated from the Robie and Waldbaum tabulations (41.59
e.u. at 800 K and 40.16 e.u. at 1000 K);   but both the
latter are based on the assumption of complete Al-Si
disorder in sanidine. Since Chatterjee and Johannes
found a certain amount of Al-Si order in their product
sanidine, the ΔS_T^0 of the reaction actually studied by
them is likely to be slightly smaller than that obtained
from Robie and Waldbaum.

Adopting ΔS_T^0 = + 38.0 e.u., one obtains for $\Delta H_{1,T}^0$
21 196 cal < $H_{1,T}^0$ < 21 343 cal. If ΔS^0 were to be fixed
at + 37.0 e.u., slightly lower values of $\Delta H_{1,T}^0$ are
obtained, viz. 20 249 cal < $H_{1,T}^0$ < 20 418 cal.

2.   Reaction (1) has an enthalpy change at 1 bar and any
temperature, $\Delta H_{1,T}^0$, given by

$$\Delta H_{1,T}^0 = (H_{and}^0)_{1,T} + (H_{san}^0)_{1,T} + (H_{steam}^0)_{1,T} -$$

$$- (H_{musc}^0)_{1,T} - (H_{qz}^0)_{1,T}.$$

The temperature 900 K is approximately in the middle
of the experimental temperature range investigated by
Chatterjee and Johannes so that $\Delta H_{1,298}^0$ can be obtained
from $\Delta H_{1,900}^0$ using the values of $(H_{900} - H_{298})$ given :

$$\Delta H_{900}^0 - \Delta H_{1,298}^0 = (H_{900} - H_{298})_{and} + (H_{900} - H_{298})_{san} +$$

$$+ (H_{900} - H_{298})_{steam} - (H_{900} - H_{298})_{musc} -$$

$$- (H_{900} - H_{298})_{qz}$$

$$= - 2860 \text{ cal.}$$

$$\Delta H^0_{1,298} = \Delta H^0_{1,900} + 2860 \text{ cal.}$$

Using, for example, the value of $\Delta H^0_{1,T}$ derived by assuming $\Delta S^0_T = + 38.0$ e.u.,

$$\Delta H^0_{1,298} = 21\ 270 + 2860 = 24\ 130 \text{ cal.}$$

$\Delta H^0_{1,298}$ may also be obtained from enthalpy of formation data as follows :

$$\Delta H^0_{1,298} = (H_f^{and})_{1,298} + (H_f^{san})_{1,298} + (H_f^{steam})_{1,298}$$

$$- (H_f^{musc})_{1,298} - (H_f^{qz})_{1,298}$$

Substituting for $\Delta H^0_{1,298}$ and all H_f except muscovite :

$$24\ 130 = - 619\ 390 - 944\ 378 - 57\ 796 - (H_f^{musc})_{1,298} +$$

$$+ 217\ 650 \text{ cal}$$

$$(H_f^{musc})_{1,298} = \underline{- 1\ 428\ 044 \text{ cal}}$$

Note that this value of $(H_f^{musc})_{1,298}$ is not consistent with all the entropy data in Robie and Waldbaum since it has been derived by assuming $\Delta S^0_T = + 38.0$ e.u. at 900 K.

Appendix: The thermodynamic properties of gases

1. INTRODUCTION

If the chemical potential of the component i of a pure phase j is known at some pressure P^0 and temperature T, μ_i^j can be calculated at any other pressure P and the same temperature T from :

$$(\mu_i^j)_{P,T} = (G_j)_{P,T} = (G_j)_{P^0,T} + \int_{P^0}^{P} V_j \, dP . \qquad (A.1)$$

If phase j is a solid, then V_j may often be taken to be independent of pressure so that integration of (A.1) gives :

$$(\mu_i^j) = (G_j)_{P,T} = (G_j)_{1,T} + (P-P^0) \, V_j . \qquad (A.2)$$

If the phase is a real gas, however, V_j is a complex function of pressure and integration of eqn (A.1) is usually rather tedious. For many gases of geological interest, however, it is not necessary to perform these integrations oneself because the results are already tabulated in the literature. The results have been derived from measured pressure-volume-temperatures (PVT) relationships of the pure gases and are usually presented in terms of fugacity f_i or fugacity coefficient Γ_i. The fugacity of f_i of a pure gas i at pressure P is defined as follows :

$$RT \ln f_i = RT \ln P + \int_0^P (V_i - V_p) \, dP. \qquad (A.3)$$

where V_i is the molar volume of i and V_p is the molar volume of a perfect gas under the same conditions (RT/P). It is obvious from (A.3) that, if the gas is perfect, the result given in Chapter 2 is obtained :

$$RT \ln f_i = RT \ln P$$

If V_i is different from the volume of a perfect gas, then f_i is not equal to P.   In this case it is necessary to introduce the fugacity coefficient Γ_i defined as follows :

$$f_i = P_i \, \Gamma_i. \qquad (A.4)$$

Comparison of (A.4) with (A.3) gives

$$RT \ln \Gamma_i = \int_0^P (V_i - V_p) \, dP. \qquad (A.5)$$

The integral on the right-hand side of eqn (A.1) (i.e. $(\mu_i^j)_{P,T} - (\mu_i^j)_{P^0,T}$) may be expressed in terms of f_i and

f_i^0 as follows

$$RT \ln f_i - RT \ln f_i^0 = RT \ln P - RT \ln P^0 + \int_0^P (V_i - V_p) \, dP -$$

$$- \int_0^{P^0} (V_i - V_p) \, dP. \qquad (A.6)$$

Rearranging and collecting terms in V_i and $\bar{V}_p$:

$$RT \ln \frac{f_i}{f_i^0} = RT \ln \frac{P}{P^0} + \int_P^{P^0} V_p \, dP + \int_{P^0}^P V_i \, dP. \qquad (A.7)$$

Substituting $V_p = RT/P$ and integrating the second term on the right-hand side of (A.7) gives

$$RT \ln \frac{f_i}{f_i^0} = RT \ln \frac{P}{P^0} + RT \ln \frac{P^0}{P} + \int_0^P V_i \, dP$$

$$\qquad (A.8)$$

$$RT \ln \frac{f_i}{f_i^0} = \int_{P^0}^P V_i \, dP.$$

Thus, the integral on the right-hand side of (A.1) may be obtained from tabulations of fugacity as a function of pressure and temperature. Since fugacity is related to pressure (eqn (A.4)), eqn (A.8) becomes

$$RT \ln \frac{P \Gamma_i}{P^0 \Gamma_i^0} = \int_{P^0}^P V_i \, dP. \qquad (A.9)$$

At very low pressures, intermolecular forces are small because the average intermolecular distances are large. Thus, at low pressures, real gas behaviour approaches that of perfect gases (the latter having no intermolecular forces) and V_i approaches V_p. If, therefore, the standard thermodynamic data are taken at very low pressure, the standard state fugacity coefficient for i, Γ_i^0, may be taken as equal to 1.0.

2. H_2O

Burnham et al. (1969) have evaluated the integrals on the right-hand side of eqn (A.9) for H_2O using a reference pressure of 0.01 bar. From these integrals they tabulate fugacity and fugacity coefficients at P and T from 20 to 1000°C and 100 to 10 000 bars. Their tables are based on the assumption that H_2O behaves as a perfect gas at 0.01 bar. The fugacity given in Table 5 of their work is therefore :

$$f_{H_2O}^* = \frac{P\,\Gamma_{H_2O}^P}{\Gamma_{H_2O}^{0.01}}.$$

Assuming that $\Gamma_{H_2O}^{0.01}$ is equal to 1.0, their tabulated values of $f_{H_2O}^*$ are equal to f_{H_2O}. Similarly, the fugacity coefficients given in Table 6 of Burnham et al. are :

$$\Gamma^*_{H_2O} \;=\; \frac{\Gamma^P_{H_2O}}{\Gamma^{0.01}_{H_2O}} \; .$$

Since the assumption that $\Gamma^{0.01}_{H_2O}$ is equal to 1.0 is adequate for all geological purposes,[2] the fugacities and fugacity coefficients given by Burnham et al. may be used directly for $f^P_{H_2O}$ and $\Gamma^P_{H_2O}$

3. CALCULATION OF $\mu^P_{H_2O}$ FROM THE DATA OF BURNHAM ET AL. (1969)

The free energy and enthalpy data given in Burnham et al. are based on unusual standard states for H_2O (see pages 7-15 of their work) which cannot be readily used in conjunction with the 1 atmosphere enthalpies of formation from the elements which are tabulated in Robie and Waldbaum (1968).   The authors consider it easiest to use the tabulated enthalpy of formation data from Robie and Waldbaum and to extrapolate these to higher pressures using the fugacity data of Burnham et al. or of Helgeson and Kirkham (1974) (Table A.1).

It should be noted that the free energy and enthalpy data given by Robie and Waldbaum refer to 1 atmosphere (1.01325 bar) not to 1 bar.   In order to correct these to a 1 bar reference state, the following addition should strictly be made :

$$(G_f)_{1 \text{ bar},T} = (G_f)_{1 \text{ atm},T} + 0.01308T \text{ cal}$$

(in many cases the additional term is insignificantly small). Given $(\mu_{H_2O})_{1\,bar,T}$ for pure H_2O steam, the chemical potential at any other pressure P is simply given by

$$(\mu_{H_2O})_{P,T} = (\mu_{H_2O})_{1,T} + RT \ln \left(\frac{P\Gamma_{H_2O}^{P}}{1.\Gamma_{H_2O}^{1\,bar}} \right) \tag{A.10}$$

where $\Gamma_{H_2O}^{P}$ is taken from Burnham et al. or Helgeson and Kirkham. Since the Robie and Waldbaum data refer to a hypothetical state of perfect H_2O gas at 1 atm, then $\Gamma_{H_2O}^{1\,atm}$ (or $\Gamma_{H_2O}^{1\,bar}$) using their data is by definition equal to 1.0. Hence $(\mu_{H_2O})_{P,T}$ may be calculated.

The activity of pure H_2O at P bar and T, using a standard state of a hypothetical ideal gas at 1 bar, is given by

$$(\mu_{H_2O})_{P,T} = (\mu_{H_2O}^{0})_{1,T} + RT \ln a_{H_2O}$$

where

$$a_{H_2O} = \frac{P\Gamma_{H_2O}^{P}}{1.1} = P\Gamma_{H_2O}^{P}.$$

If the H_2O component is present in a mixed fluid of mole fraction X_{H_2O} of H_2O, and if mixing of gases is ideal (Chapter 3), the activity relative to the 1 bar standard state of hypothetical ideal gas is

$$a_{H_2O} = P\Gamma^P_{H_2O} x_{H_2O}.$$

Except at very low temperatures ($< 200^\circ C$), the actual 1 bar thermodynamic properties of H_2O gas are effectively the same as those of the hypothetical gas whose properties are tabulated by Robie and Waldbaum.

4. OTHER GASES

Estimated fugacity coefficients for CO_2 are tabulated at pressures of 500 to 10 000 atmospheres by Mel'nik (1972) and for 500 to 3000 bar by Skippen (1971) (Table A.2). Skippen also gives data for CO and CH_4 at low pressures. These fugacity coefficients may be used in conjunction with Robie and Waldbaum or JANAF tabulations to calculate μ_i at high pressures. Fugacity coefficients for hydrogen at pressures up to 3 kbar and temperatures of 0 to $1000^\circ C$ are given by Shaw and Wones (1964).

Apart from experiments on H_2O, there has been little work on the PVT relationship of gases at pressures above 3 kbar. It is, however, possible to estimate fugacity coefficients for imperfect gases at higher pressures using the principle of corresponding states.

It is assumed that the properties of all real gases are similar functions of pressure and temperature when these properties are calculated relative to the critical pressure, critical temperature, and critical volume of the gas. Thus, the reduced pressure P_r, reduced temperature T_r, and reduced volume V_r of an imperfect gas are defined as follows :

$$P_r = \frac{P}{P_c} \quad (P_c = \text{critical pressure})$$

$$T_r = \frac{T}{T_c} \quad (T_c = \text{critical temperature})$$

$$V_r = \frac{V}{V_c} \quad (V_c = \text{critical volume}).$$

Mel'nik (1972) has plotted values of $\log \Gamma_i$ versus P/P_c at different values of T/T_c for a number of real gases. Extremely good correlations were found (see Newton 1935). Thus, Γ_i for gases which have not been studied at high pressures may be calculated from experimental results on other gases using correlations based on P_r, T_r, and V_r.

TABLE A.1

Values of Γ_{H_2O}

T°_C					P(kbar)					
	0.5	1	2	3	4	5	6	7	8	9
200	0.0366	0.0236	0.019	0.020	0.024	0.030	0.038	0.050	0.066	0.089
250	0.0859	0.0546	0.043	0.045	0.051	0.062	0.077	0.097	0.125	0.162
300	0.1659	0.1050	0.082	0.083	0.092	0.109	0.132	0.163	0.205	0.259
350	0.2763	0.1752	0.135	0.135	0.148	0.171	0.203	0.246	0.303	0.375
400	0.4083	0.2622	0.202	0.199	0.216	0.245	0.287	0.343	0.414	0.504
450	0.5416	0.3594	0.278	0.273	0.293	0.329	0.380	0.447	0.533	0.640
500	0.6481	0.4589	0.361	0.353	0.376	0.418	0.478	0.556	0.654	0.776
550	0.7259	0.5528	0.445	0.435	0.461	0.510	0.577	0.665	0.775	0.909
600	0.7848	0.6358	0.527	0.517	0.546	0.600	0.675	0.771	0.890	1.034
650	0.8297	0.7057	0.603	0.595	0.628	0.686	0.767	0.870	0.997	1.150
700	0.8646	0.7633	0.673	0.668	0.704	0.767	0.853	0.962	1.095	1.254
800	0.9137	0.8487	0.789	0.795	0.839	0.910	1.004	1.122	1.263	1.429

Taken from Helgeson and Kirkham (1974).   Standard state of perfect gas at 1 bar.

TABLE A.2

Values of Γ_{CO_2} from Mel'nik (1972)

P(atm)	\multicolumn{12}{c}{T(°K)}
	400	500	600	700	800	900	1000	1100	1200	1300	1400	1500
500	0.57	0.85	1.02	1.11	1.15	1.17	1.19	1.20	1.20	1.20	1.20	1.20
1000	0.63	0.95	1.16	1.28	1.34	1.37	1.38	1.38	1.37	1.37	1.37	1.37
1500	0.82	1.20	1.44	1.56	1.62	1.62	1.62	1.59	1.59	1.57	1.55	1.53
2000	1.16	1.60	1.85	1.95	2.00	1.95	1.91	1.86	1.82	1.78	1.76	1.71
3000	2.46	3.00	3.18	3.15	2.95	2.08	2.46	2.34	2.29	2.19	2.09	2.04
4000	5.49	5.60	5.60	5.18	4.67	3.89	3.55	3.31	3.16	2.95	2.75	2.63
5000	12.5	11.3	9.88	8.52	6.92	5.75	5.01	4.57	4.27	3.89	3.63	3.39
6000	28.4	22.1	17.4	13.9	10.5	8.32	7.24	6.31	5.75	5.13	4.68	4.27
7000	64.2	42.7	30.2	22.4	16.2	12.6	10.5	8.91	7.94	6.92	6.17	5.50
8000	143	81.4	51.8	35.7	24.6	18.6	15.1	12.6	10.7	9.12	7.94	7.08
9000	319	154	88.1	56.3	38.0	28.8	22.4	17.8	14.8	12.3	10.5	9.33
10000	703	289	148	80.3	56.2	41.7	31.6	24.6	20.0	16.2	13.5	11.8

Pressure in atmospheres. Derived from eqns (A.3) and (A.4). Using Robie and Waldbaum data, $\Gamma_{CO_2}^{1\ bar} = 1$.

References

ALBAREDE, F. and BOTTINGA, Y. (1972). Kinetic disequilibrium in trace element partitioning between phenocrysts and host lava. Geochim. Cosmochim. Acta 36, 141-56.

ALTHAUS, E., KAROTKE, E., NITSCH, K.H., WINKLER, H.G.F. (1970). An experimental re-examination of the upper stability limit of muscovite plus quartz. Neues. Jb. Miner. Mh. 1970, 325-36.

ANDERSON, G.M. (1970). Some thermodynamics of dehydration equilibria. Am. J. Sci. 269, 392-401.

ARTH, J.G. (1976). Behaviour of trace elements during magmatic processes - a summary of theoretical models and their applications. J. Res. U.S. Geol. Survey 4, 41-7.

BACON, C.R. and CARMICHAEL, I.S.E. (1973). Stages in the P-T path of ascending basalt magma: an example from San Quentin, Baja California. Contr. Miner. Petrology 41, 1-22.

BANNO, S. (1970). Classification of eclogites in terms of physical conditions of their origin. Phys. Earth & Planet. Interiors 3, 405-21.

BARRON, L.M. (1973). Nonideal thermodynamic properties of H_2O-CO_2 mixtures for 0.4-2 Kb and 400-700°C. Contr. Mineral. Petrology 39, 184.

BELL, P.M. (1963). Aluminium silicate systems: experimental determination of the triple point. Science N.Y. 139, 1055-56.

BELTON, G.R., SUITO, H. and GASKELL, D.R. (1973). Free energies of mixing in the liquid iron-cobalt ortho-silicates at 1450°C. Met. Trans. 4, 2541-7.

BIGGAR, G.M. and O'HARA, M.J. (1969). Solid solutions at atmospheric pressure in the system $CaO-MgO-SiO_2$ with special reference to the instabilities of diopside, akermanite and monticellite. Progress in Experimental Petrology N.E.R.C., 1st Report, 89-96.

BIRCH, F., SCHAIRER, J.F. and SPICER, H.C. (1942). Handbook of physical constants. Geol. Soc. Am. Spec. Paper 36, 325.

BOCKRIS, J. O'M, KITCHENER, J.A. and DAVIES, A.E. (1952). Electric transport in liquid silicates. Trans. Faraday Soc. 48, 536-48.

BOETTCHER, A.L. (1970). The system $CaO-Al_2O_3-SiO_2-H_2O$ at high pressures and temperatures. J. Petrology 11, 337-79.

BOWEN, N.L. (1928). The evolution of the igneous rocks. Princeton University Press, Princeton, N.J.

BOWEN, N.L. and SCHAIRER, J.F. (1935). The system MgO-FeO-SiO_2. Am. J. Sci. 5th Ser., 29, 151-217.

BOYD, F.R. and ENGLAND, J.L. (1963). Effect of pressure on the melting points of diopside, $CaMgSi_2O_6$, and albite $NaAlSi_3O_8$, in the range up to 50 kilobars. J. geophys. Res. 68, 311-23.

BOYD, F.R., ENGLAND, J.L. and DAVIS, B.T.C. (1964). Effects of
pressure on the melting and polymorphism of enstatite,
$MgSiO_3$. J. geophys. Res. 69, 2101-9.

BUDDINGTON, A.F. and LINDSLEY, D.H. (1964). Iron-titanium
oxide minerals and synthetic equivalents. J. Petrology 5
310-57.

BURNHAM, C.W. (1975). Water and magmas: a mixing model.
Geochim. Cosmochim. Acta 39, 1077-84.

BURNHAM, C.W. and DAVIS, N.F. (1974). The role of H_2O in
silicate melts, II. Thermodynamic and phase relations
in the system $NaAlSi_3O_8$-H_2O to 10 kilobars, 700° to
$1100^o C$. Am. J. Sci. 274, 902-40.

BURNHAM, C.W., HOLLOWAY, J.R. and DAVIS, N.F. (1969). Thermo-
.dynamic properties of water to $1000^o C$ and 10,000 bars.
Geol. Soc. Am. Spec. Paper 132, 1-96.

CARMICHAEL, I.S.E. (1967a). The mineralogy of Thingmuli, a
tertiary volcano in eastern Iceland. Am. Miner. 52,
1815-41.

CARMICHAEL, I.S.E. (1967b). The iron-titanium oxides of salic
volcanic rocks and their associated ferromagnesian
silicates. Contr. Miner. Petrology 14, 36-64.

CARMICHAEL, I.S.E. (1967c). The mineralogy and petrology of
the volcanic rocks from the Leucite Hills, Wyoming.
Contr. Miner. Petrology 15, 24-66.

CHATTERJEE, N.D. and JOHANNES, W. (1974). Thermal stability
and standard thermodynamics of synthetic "M_1 muscovite"
$KAl_2\{AlSi_3O_{10}(OH)_2\}$. Contr. Miner. Petrology 48, 89-114.

CLARK, J.R., APPLEMAN, D.E. and PAPIKE, J.J. (1969). Crystal-chemical characterization of clinopyroxenes based on eight new structure referents. Mineralog. Soc. Am. Spec. Paper 2, 31-50.

CLARK, S.P. (1966). Handbook of physical constants. Geol. Soc. Am. Mem. 97.

DAVIS, B.T.C. and ENGLAND, J.L. (1964). The melting of forsterite up to 50 kilobars. J. geophys. Res. 69, 1113-6.

DAY, H.W. (1973). The temperature stability of muscovite plus quartz. Am. Miner. 58, 255-62.

DEER, W.A., HOWIE, R.A. and ZUSSMAN, J. (1962). Rock forming minerals. Wiley, New York.

DRAKE, M.J. (1975). The oxidation state of europium as an indicator of oxygen fugacity. Geochim. Cosmochim. Acta 39, 55-64.

DRAKE, M.J. and WEILL, D.F. (1975). The partition of Sr, Ba, Ca, Y, Eu^{2+}, Eu^{3+}, and other REE between plagioclase feldspar and magmatic silicate liquid: an experimental study. Geochim. Cosmochim. Acta 39, 689-712.

EGGLER, D.H. (1973). Role of CO_2 in melting processes in the mantle. Yb. Carnegie Instn. Wash. 72, 457-67.

EGGLER, D.H. (1976). Does CO_2 cause partial melting in the low velocity layer in the mantle? Geology 4, 69-72.

ESIN, O.A. (1973). Ideal ionic solutions of silicate polymers. Russ. J. phys. Chem. 47, 1306-8.

ESKOLA, P. (1920). The mineral facies of rocks. Norsk geol. Tidsskr. 6, 143-94.

EUGSTER, H.P., ALBEE, A.L., BENCE, A.E., THOMPSON, J.B. Jr. and WALDBAUM, D.R. (1972). The two-phase region and

excess mixing properties of paragonite-muscovite
crystalline solutions. J. Petrology 13, 147-79.

EUGSTER, H.P. and SKIPPEN, G.B. (1967). Igneous and metamorphic reactions involving gas equilibria. In Researches
in geochemistry (ed. P.H. Abelson), 2, 492-520. Wiley,
New York.

EVANS, B.W. (1965). Application of a reaction rate method
to the breakdown equilibria of muscovite and muscovite
plus quartz. Am. J. Sci. 263, 647-67.

EVANS, B.W. and WRIGHT, T.L. (1972). Composition of liquidus
chromite from the 1959 (Kilauea Iki) and 1965 (Makaopuhi)
eruptions of Kilauea Volcano, Hawaii. Am. Miner. 57,
217-30.

FINCHAM, C.J.B. and RICHARDSON, R.F. (1954). The behaviour
of sulphur in silicate and aluminate melts. Proc. Roy.
Soc. A223, 40-61.

FISHER, J.R., ZEN, E-AN (1971). Thermochemical calculations
from hydrothermal phase equilibrium data and the free
energy of H_2O. Am. J. Sci. 270, 297-314.

FRASER, D.G. (1975a). Activities of trace elements in silicate
melts. Geochim. Cosmochim. Acta 39, 1525-30.

FRASER, D.G. (1975b). An investigation of some long-chain
oxy-acid systems. D. Phil. Thesis University of Oxford.

GANGULY, J. (1973). Activity-composition relation of jadeite
in omphacite pyroxene. Earth Planet. Sci. Lett. 19,
145-53.

GANGULY, J. and KENNEDY, G.C. (1974). The energetics of garnet
solid solution. I. Mixing of the alumino-silicate end
members. Contr. Miner. Petrology 48, 137.

GAST, P.W. (1968). Trace element fractionation and the origin
    of tholeiitic and alkaline magma types. Geochim. Cosmochim.
    Acta 32, 1057-86.

GORDON, T.M. (1973). Determination of internally consistent
    thermodynamic data from phase equilibrium experiments.
    J. Geol. 81, 199-208.

GREENLAND, L.P. (1970). An equation for trace element
    distribution during magmatic crystallization. Am. Miner.
    55, 455-65.

GREENWOOD, H.J. (1963). The synthesis and stability of
    anthophyllite. J. Petrol. 4, 315-35.

GREENWOOD, H.J. (1967). Wollastonite: stability in H_2O-CO_2
    mixtures and occurrence in a contact-metamorphic
    aureole near Salmo, British Columbia, Canada. Am. Miner.
    52, 1669-80.

GREENWOOD, H.J. (1973). Thermodynamic properties of gaseous
    mixtures of H_2O and CO_2 between 450° and $800^\circ C$ and 0 to
    500 bar. Am. J. Sci. 273, 561-71.

GRUTZECK, M.W., KRIDELBAUGH, S.J. and WEILL, D.F. (1973).
    REE partitioning between diopside and silicate liquid.
    EOS 54, 1222.

GUGGENHEIM, E.A. (1952). Mixtures. Clarendon Press, Oxford.

HÄKLI, T.A. and WRIGHT, T.L. (1967). The fractionation of
    nickel between olivine and augite as a geothermometer.
    Geochim. Cosmochim. Acta 31, 877-84.

HARIYA, Y. and KENNEDY, G.C. (1968). Equilibrium study of
    anorthite under high pressure and high temperature.
    Am. J. Sci. 266, 193-203.

HAYS, J.F. (1966). Lime-alumina-silica. Yb. Carnegie Instn.
    Wash. 65, 234-9.

HELGESON, H.C. and KIRKHAM, D.H. (1974). Theoretical
    prediction of the thermodynamic behaviour of aqueous
    electrolytes at high pressures and temperatures:
    I Summary of the thermodynamic/electrostatic properties
    of the solvent. Am. J. Sci. 274, 1089-1198.

HENDERSON, L.M. and KRACEK, F.C. (1927). The fractional
    precipitation of barium and radium chromates. J. Am.
    Chem. Soc. 49, 739-49.

HENSEN, B.J., SCHMID, R. and WOOD, B.J. (1975). Activity-
    composition relations in grossular-pyrope solid solutions.
    Contr. Miner. Petrology 51, 161-66.

HERTOGEN, J. and GIJBELS, R. (1976). Calculations of trace
    element fractionation during partial melting. Geochim.
    Cosmochim. Acta 40, 313-22.

HOLDAWAY, M.J. (1971). Stability of andalusite and the
    aluminium silicate phase diagram. Am. J. Sci. 271,
    97-131.

IRVINE, T.N. (1965). Chromian spinel as a petrogenetic
    indicator: Part 1, Theory. Can. J. Earth Sci. 2, 648-72.

IRVINE, T.N. (1975). Crystallization sequences in the
    Muskox intrusion and other layered intrusions - II.
    Origin of chromitite layers and similar deposits of
    other magmatic ores. Geochim. Cosmochim. Acta 39, 991-
    1020.

JACKSON, E.D. (1969). Chemical variation in coexisting
    chromite and olivine in chromite zones of the Stillwater
    complex. Econ. Geol. Mon. 4, 41-71.

JANAF (1971). Thermochemical tables (by D.R. Stull and others). National Bureau of Standards. National Reference Data System. NSRDS-NB537.

JONES, J.W. (1972). An almandine garnet isograd in the Rogers Pass area, British Columbia. The nature of the reaction and an estimation of the physical conditions during its formation. Contr. Miner. Petrology 37, 291-306.

KAY, R.W. & GAST, P.W. (1973). The rare earth content and origin of alkali basalts. J. Geol. 81, 653-82.

KELLEY, K.K. (1960). Contributions to the data on theoretical metallurgy: pt. 13, high temperature heat content, heat capacity and entropy data for the elements and inorganic compounds. U.S. Bur. Mines Bull. 584.

KING, M.B. (1969). Phase equilibrium in mixtures. Pergamon Press, Oxford.

KITAYAMA, K. and KATSURA, T. (1968). Activity measurements in orthosilicate and metasilicate solid solutions. I. Mg_2SiO_4-Fe_2SiO_4 and $MgSiO_3$-$FeSiO_3$ at 1204^oC. J. chem. Soc. Japan 41, 1146-51.

KUBASCHEWSKI, O., EVANS, B.W. and ALCOCK, C.B. (1967). Metallurgical thermochemistry. Pergamon Press, Oxford.

KURKJIAN, C.R. and RUSSELL, L.E. (1958). Solubility of water in molten alkali silicates. J. Soc. Glass Technol. 42, 130.

KUSHIRO, I. (1969). The system forsterite-diopside-silica with and without water at high pressures. Am. J. Sci. Schairer Vol., 267A, 269-94.

KUSHIRO, I. (1972). Effect of water on the composition of magmas formed at high pressures. J. Petrology 13, 311-34.

KUSHIRO, I. (1973). The system diopside-anorthite-albite: determination of compositions of coexisting phases. Yb. Carnegie Instn. Wash. 72, 502-7.

KUSHIRO, I. (1975). On the nature of silicate melt and its significance in magma genesis. Regularities in the shift of the liquidus boundaries involving olivine, pyroxene and silica minerals. Am. J. Sci. 275, 411-31.

LEEMAN, W.P. (1973). Partitioning of Ni and Co between olivine and basaltic liquid: an experimental study. EOS 54, 1222.

LINDSLEY, D.H. (1962). Investigations in the system FeO-Fe_2O_3-TiO_2. Yb. Carnegie Instn. Wash. 61, 100-6.

LINDSLEY, D.H. (1963). Equilibrium relations of coexisting pairs of Fe-Ti oxides. Yb. Carnegie Instn. Wash. 62, 60-6.

LONEY, R.A., HIMMELBERG, G.R. and COLEMAN, R.G. (1971). Structure and petrology of the Alpine-type peridotite at Burro Mountain, California, U.S.A. J. Petrology 12, 245-309.

MACGREGOR, I.D. (1974). The system MgO-Al_2O_3-SiO_2: solubility of Al_2O_3 in enstatite for spinel and garnet peridotite compositions. Am. Miner. 59, 110-19.

MCINTIRE, W.L. (1963). Trace element partition coefficients - a review of theory and applications to geology. Geochim. Cosmochim. Acta 27, 1209-64.

MASSON, C.R. (1965). An approach to the problem of ionic distribution in liquid silicates. Proc. Roy. Soc. A287, 201-21.

MASSON, C.R., SMITH, I.B. and WHITEWAY, S.G. (1970)
Activities and ionic distribution in liquid silicates :
application of polymer theory. Can. J. Chem. 48, 1456-64.

MEL'NIK, Y.P. (1972). Thermodynamic parameters of
compressed gases and metamorphic reactions involving water
and carbon dioxide. Geochem. Int. 9, 419-26.

MORRIS, R.V. and HASKIN, L.A. (1974). EPR measurement of the
effect of glass composition on the oxidation states of
europium. Geochim. Cosmochim. Acta 38, 1435-45.

MYSEN, B.O., and BOETTCHER, A.L. (1975). Melting of a hydrous
mantle. I. Phase relations of natural peridotite at high
pressures and temperatures with controlled activities of
water, carbon dioxide and hydrogen. J. Petrology 16,
520-48.

NAFZIGER, R.H. (1973). High-temperature activity-composition
relations of equilibrium spinels, olivines and pyroxenes
in the system Mg-Fe-O-SiO$_2$. Am. Miner. 58, 457-65.

NAFZIGER, R.H. and MUAN, A. (1967). Equilibrium phase
compositions and thermodynamic properties of olivines and
pyroxenes in the system MgO-FeO-SiO$_2$. Am. Miner. 52,
1364-85.

NERNST, W. (1891). Verteilung eines Stoffes zwischen zwei
Lösungsmitteln und zwischen Lösungsmitteln und Dampfraum.
Z. Phys. Chem. 8, 110.

NEWTON, R.C. (1966). Some calc-silicate equilibrium relations.
Am. J. Sci. 264, 204-22.

NEWTON, R.C. (1969). Some high-pressure hydrothermal experiments on severely ground kyanite and sillimanite. Am. J.
Sci. 267, 278-284.

NEWTON, R.H. (1935). Activity coefficients of gases. Ind. Engng Chem. ind. Edn. 27, 302-6.

NICHOLLS, J. and CARMICHAEL, I.S.E. (1972). The equilibration temperature and pressure of various lava types with spinel- and garnet-peridotite. Am. Miner. 57, 941-59.

O'HARA, M.J. (1968). The bearing of phase equilibria studies in synthetic and natural systems on the origin and evolution of basic and ultrabasic rocks. Earth-sci. Rev. 4, 69-133.

O'NIONS, R.K. and PANKHURST, R.J. (1974). Petrogenetic significance of isotope and trace element variations in volcanic rocks from the Mid-Atlantic. J. Petrology 15, 603-34.

ORR, R.L. (1953). High temperature heat contents of magnesium orthosilicate and ferrous orthosilicate. J. Am. chem. Soc. 75, 528-9.

ORVILLE, P.M. (1972). Plagioclase cation exchange equilibria with aqueous chloride solution at $700^{\circ}C$ and 2000 bars in the presence of quartz. Am. J. Sci. 272, 234-72.

OSBORN, E.F. (1942). The system $CaSiO_3$-diopside-anorthite. Am. J. Sci. 240, 751-88.

PAUL, A. and DOUGLAS, R.W. (1965). Ferrous-ferric equilibrium in binary alkali silicate glasses. Physics Chem. Glasses 6, 207-11.

POWELL, R. (1974). A comparison of some mixing models for crystalline silicate solid solutions. Contr. Miner. Petrology 46, 265-74.

RAMBERG, H. and DE VORE, G.W. (1951). Distribution of Fe^{2+} and Mg^{2+} in coexisting olivines and pyroxenes. J. Geol.

59, 193-210.

RAYLEIGH, J.W.S. (1896). Theoretical considerations respecting the separation of gases by diffusion and similar processes. Phil. Mag. 42, 77-107.

RICHARDSON, F.D. (1956). Activities in ternary silicate melts. Trans. Faraday Soc. 52, 1312-24.

RICHARDSON, S.W., GILBERT, M.C. and BELL, P.M. (1969). Experimental determination of kyanite-andalusite and andalusite-sillimanite equilibria; the aluminium silicate triple point. Am. J. Sci. 267, 259-272.

RINGWOOD, A.E. (1956). Melting relationships of Ni-Mg olivines and some geochemical implications. Geochim. Cosmochim. Acta 10, 297-303.

ROBIE, R.A. and WALDBAUM, D.R. (1968). Thermodynamic properties of minerals and related substances at $298.15^{\circ}K$ $(25.0^{\circ}C)$ and one atmosphere (1.013 bars) pressure and at higher temperatures. Bull. U.S. geol. Surv. 1259.

ROEDDER, E. (1972). Composition of fluid inclusions. Data of geochemistry, 6/e U.S. geol. Surv. Prof. Paper, 440, JJ.

ROEDER, P.L. and EMSLIE, R.F. (1970). Olivine-liquid equilibrium. Contr. Miner. Petrology 29, 275-89.

RUMBLE, D., III (1970). Thermodynamic analysis of phase equilibria in the system Fe_2TiO_4-Fe_3O_4-TiO_2. Yb. Carnegie Instn. Wash. 69, 198-206.

RUMBLE, D., III (1973). Oxide minerals from regionally metamorphosed quartzites of Western New Hampshire. Contr. Miner. Petrology 42, 181-95.

SAXENA, S.K. (1973). Crystalline solutions. Springer,
Berlin.

SAXENA, S.K. and GHOSE, S. (1971). $Mg^{2+}-Fe^{2+}$ order-disorder
and the thermodynamics of the orthopyroxene-crystalline
solution. Am. Miner. 56, 532-59.

SAXENA, S.K. and RIBBE, P.H. (1972). Activity-composition
relations in feldspars. Contr. Miner. Petrology 37, 131-8.

SCARFE, C.M., LUTH, W.C. and TUTTLE, O.F. (1966). An
experimental study bearing on the absence of leucite in
plutonic rocks. Am. Miner. 51, 726-35.

SCHAIRER, J.F. and YODER, H.S., Jr. (1960). The nature of
residual liquids from crystallization, with data on the
system nepheline-diopside-silica. Am. J. Sci. Bradley
Vol. 258A, 273-83.

SCHILLING, J.G. and WINCHESTER, J.W. (1967). Rare earth
fractionation and magmatic processes. In   Mantles of
the earth and terrestrial planets (ed. S.K. Runcorn).
Interscience, London.

SCHNETZLER, C.C. and PHILPOTTS, J.A. (1970). Partition
coefficients of rare-earth elements between igneous
matrix material and rock-forming mineral phenocrysts -
II. Geochim. Cosmochim. Acta 34, 331-40.

SCHWERDTFEGER, K., MUAN, A. and DARKEN, L.S. (1966).
Activities in olivine and pyroxenoid solid solutions of
the system Fe-Mn-Si-O. Trans. Metall. Soc. A.I.M.E.
236, 201-11.

SHAW, D.M. (1970). Trace element fractionation during
anatexis. Geochim. Cosmochim. Acta 34, 237-43.

SHAW, H.R. and WONES, D.R. (1964). Fugacity coefficients
   for hydrogen gas between 0° and $1000^{\circ}C$ for pressures to
   3000 atm. Am. J. Sci. 262, 918-29.

SHIMIZU, N. and KUSHIRO, I. (1975). The partitioning of rare
   earth elements between garnet and liquid at high
   pressures: preliminary experiments. Geophys. Res. Lett.
   2, 413-16.

SKIPPEN, G.B. (1971). Experimental data for reactions in
   siliceous marbles. J. Geol. 79, 457-81.

SKIPPEN, G.B. (1974). An experimental model for low
   pressure metamorphism of siliceous dolomitic marble.
   Am. J. Sci. 274, 487-509.

SUN, C.O., WILLIAMS, R.J. and SUN, S.S. (1974). Distribution
   coefficients of Eu and Sr for plagioclase-liquid and
   clinopyroxene-liquid equilibria in oceanic ridge basalt :
   an experimental study. Geochim. Cosmochim. Acta 38,
   1415-33.

SUN, S.S. and HANSON, G.N. (1975). Origin of Ross Island
   basanitoids and limitations upon the heterogeneities of
   mantle sources for alkali basalts and nephelinites.
   Contr. Miner. Petrology 52, 77-106.

TEMKIN, M. (1945). Mixtures of fused salts as ionic solutions.
   Acta Phys.-chim. U.R.S.S. 20, 411-20.

THOMPSON, A.B. (1974). Gibbs energy of aluminous minerals.
   Contr. Miner. Petrology 48, 123-36.

THOMPSON, J.B., Jr. (1967). Thermodynamic properties of
   simple solutions. In Researches in geochemistry, (ed.
   P.H. Abelson), Vol. 2, 340-61. Wiley, New York.

THOMPSON, J.B., Jr. and WALDBAUM, D.R. (1969). Mixing
    properties of sanidine crystalline solutions III.
    Calculations based on two-phase data. Am. Miner. 54,
    811-38.

TOOP, G.W. and SAMIS, C.S. (1962a). Activities of ions in
    silicate melts. Trans. Metall. Soc. A.I.M.E., 224, 878-87.

TOOP, G.W. and SAMIS, C.S. (1962b). Some new ionic concepts
    of silicate slags. Can. Met. Q. 1, 129-52.

VASLOW, F. and BOYD, G.E. (1952). Thermodynamics of
    co-precipitation: dilute solid solutions of AgBr in
    AgCl. J. Am. Chem. Soc. 74, 4691-95.

WAGER, L.R. and MITCHELL, R.L. (1951). Distribution of trace
    elements during strong fractionation of basic magma - a
    further study of the Skaergaard intrusion. Geochim.
    Cosmochim. Acta 1, 129-208.

WALDBAUM, D.R. and THOMPSON, J.B., Jr. (1968). Mixing
    properties of sanidine crystalline solutions: pt. II,
    calculations based on volume data. Am. Miner. 53,
    2000-17.

WEILL, D.F.(1966). Stability relations in the Al_2O_3-SiO_2
    system calculated from solubilities in the Al_2O_3-SiO_2 -
    Na_3AlF_6 system. Geochim. Cosmochim. Acta 30, 223-27.

WHITEWAY, S.G., SMITH, I.B. and MASSON, C.R. (1970). Theory
    of molecular size distribution in multichain polymers.
    Can. J. Chem. 48, 32-45.

WILLIAMS, R.J. (1972). Activity-composition relations in
    the fayalite-forsterite solid solution between 900° and
    1300° at low pressures. Earth Planet. Sci. Lett. 15,
    296-300.

WONES, D.R. (1972). Stability of biotite : a reply. Am. Miner. 57, 316-7.

WONES, D.R. and EUGSTER, H.P. (1965). Stability of biotite : experiment, theory and application. Am. Miner. 50, 1228-72.

WOOD, B.J. The solubility of alumina in orthopyroxene coexisting with garnet. Contrib. Miner. Petrology 46, 1-15.

WOOD, B.J. and BANNO, S. (1973). Garnet-orthopyroxene and orthopyroxene-clinopyroxene relationships in simple and complex systems. Contrib. Miner. Petrology 42, 109-24.

ZEN, E.-AN.(1971). Comments on the thermodynamic constants and hydrothermal stability relations of anthophyllite. Am. J. Sci. 270, 136-50.

ZEN, E.-AN. (1972). Gibbs free energy enthalpy and entropy of ten rock-forming minerals. Calculations, discrepancies, implications. Am. Miner. 57, 524-53.

Index